ELECTROMAGNETIC SIMULATION USING THE FDTD METHOD WITH PYTHON

ELECTROMAGNETIC SIMULATION USING THE FDTD METHOD WITH PYTHON

Third Edition

JENNIFER E. HOULE
DENNIS M. SULLIVAN

Published by John Wiley & Sons, Inc., Hoboken, New Jersey.
Published simultaneously in Canada.

Library of Congress Cataloging-in-Publication Data:

hardback: 9781119565802

Set in 10/12pt Times New Roman by SPi Global, Pondicherry, India

10 9 8 7 6 5 4 3 2 1

To Bijan
—Jennifer E. Houle

CONTENTS

ABOUT THE AUTHORS

Jennifer E. Houle received the B.S. degree in electrical engineering from the University of Idaho in 2005. Between 2005 and 2012, she was an engineer in the semiconductor industry. In 2016, she received the M.S. degree in electrical engineering from the University of Idaho and has since been active in quantum and electromagnetic simulation. She is presently the Vice President for Research at Moscow-Berlin Simulations.

Dennis M. Sullivan is a professor of Electrical and Computer Engineering at the University of Idaho. He has done extensive work in the fields of electromagnetic and quantum simulation including EM dosimetry, hyperthermia cancer therapy, nonlinear optics, and quantum semiconductors. In 1997, Dr. Sullivan received the award for the "Best Paper by a Young Investigator" from the IEEE Antennas and Propagation Society. In 2013, he was made a fellow of IEEE. He is also the author of *Quantum Mechanics for Electrical Engineers* and *Signals and Systems for Electrical Engineers I.*

PREFACE

The purpose of the third edition of this book remains the same as that of the first two editions: to enable the reader to learn the use of the finite-difference time-domain (FDTD) method in a manageable amount of time. For this reason, the first four chapters are fundamentally the same as previous editions. The major difference is the code has now been written in Python and each program contains the code for graphical outputs. The goal of these four chapters is to take the reader through one-, two-, and three-dimensional FDTD simulation and, at the same time, present the techniques for dealing with more complicated media. In addition, some basic applications of signal processing theory are explained to enhance the effectiveness of FDTD simulation.

Chapter 5 addresses some of the advantages of Python and presents some programming topics the reader may not be familiar with. Some general programming strategies and best practices are discussed, and these practices are applied to an FDTD program. Finally, an introduction to interactive widgets is presented. This is a very useful feature that can help make programs that are user-friendly to those without programming knowledge. This chapter is geared toward those who know a limited amount about Python.

Chapter 6 contains an example of a more complicated engineering project: simulating hyperthermia treatment. This is based on research done by the authors to simulate an annular phased array to plan hyperthermia cancer treatment. This chapter is meant to illustrate the power and practical application of FDTD simulations to model how a solution is obtained. The principles applied are all explained in Chapters 1–4.

Jennifer E. Houle
and Dennis M. Sullivan

GUIDE TO THE BOOK

This book has one purpose only: it enables the reader or student to learn and do three-dimensional electromagnetic simulation using the finite-difference time-domain (FDTD) method. It does not attempt to explain the theory of FDTD simulation in great detail. It is not a survey of all possible approaches to the FDTD method, nor is it a "cookbook" of applications. It is aimed at those who would like to learn and do FDTD simulation in a reasonable amount of time.

FORMAT

This book is tutorial in nature. Every chapter attempts to address an additional level of complexity. The text increases in complexity in two major ways:

Dimension of Simulation	Type of Material
One-dimensional	Free space
Two-dimensional	Complex dielectric material
Three-dimensional	Frequency-dependent material

The first section of Chapter 1 is one-dimensional simulation in free space. From there, the chapters progress to more complicated media. In Chapter 2, the simulation of frequency-dependent media is addressed. Chapter 3 introduces two-dimensional simulation, including the simulation of plane waves and how to implement the perfectly matched layer (PML). Chapter 4 introduces three-dimensional simulation.

Chapter 5 focuses on Python as an object-oriented language, coding strategies, and features to enhance FDTD simulations in the language. Chapter 6 presents a real-world application of the FDTD method and breaks down strategies for solving the problem.

SPECIFIC CHOICES DEALING WITH SOME TOPICS

There are many ways to handle individual topics having to do with FDTD simulation. This book does not attempt to address all of them. In most cases, a single approach is taken and used throughout the book for the sake of clarity. Our philosophy is that when first learning the FDTD method, it is better to learn one specific approach and learn it well, rather than to be confused by switching to different approaches. In most cases, the approach being taught is the author's own preference. This does not make it the only approach or even the best; it is just the approach that the author has found to be effective. In particular, the following are some of the choices that have been made.

1. *The Use of Normalized Units.* Maxwell's equations have been normalized by substituting

$$\widetilde{\boldsymbol{E}} = \sqrt{\frac{\varepsilon_0}{\mu_0}}\boldsymbol{E}.$$

This is a system similar to Gaussian units, which are frequently used by physicists. The reason for using it here is the simplicity in the formulation. The E and the H fields have the same order of magnitude. This has an advantage in formulating the PML, which is a crucial part of FDTD simulation.

2. *Maxwell's Equations with the Flux Density.* There is some leeway in forming the time-domain Maxwell's equations from which the FDTD formulation is developed. The following is used in Chapter 1:

$$\frac{\partial \boldsymbol{E}}{\partial t} = \frac{1}{\varepsilon_0}\nabla \times \boldsymbol{H} \tag{1}$$

$$\frac{\partial \boldsymbol{H}}{\partial t} = -\frac{1}{\mu_0}\nabla \times \boldsymbol{E}. \tag{2}$$

This is a straightforward formulation and among those commonly used. However, by Chapter 2, the following formulation using the flux density is adopted:

$$\frac{\partial \boldsymbol{D}}{\partial t} = \nabla \times \boldsymbol{H}, \tag{3}$$

$$D = \varepsilon_0 \varepsilon_r^* E, \tag{4}$$

$$\frac{\partial H}{\partial t} = -\frac{1}{\mu_0} \nabla \times E. \tag{5}$$

In this formulation, it is assumed that the materials being simulated are nonmagnetic; that is, $H = (1/\mu_0)B$. However, we will be dealing with a broad range of dielectric properties, so Eq. (4) could be a complicated convolution. There is a reason for this formulation: Eq. (3) and Eq. (5) remain the same regardless of the material; any complicated mathematics stemming from the material lies in Eq. (4). We will see that the solution of Eq. (4) can be looked upon as a digital filtering problem. In fact, the use of signal processing techniques in FDTD simulation will be a recurring theme in this book.

Z TRANSFORMS

As mentioned above, the solution of Eq. (4) for most complicated materials can be viewed as a digital filtering problem. This being the case, the most direct approach to solve the problem is to take Eq. (4) into the Z domain. Z transforms are a regular part of electrical engineering education, but not that of physicists, mathematicians, and others. In teaching a class on FDTD simulation, Prof. Sullivan teaches some Z transform theory so when he reaches the sections on complicated dispersive materials, the students are ready to apply Z transforms. This has two distinct advantages: (a) Electrical engineering students have another application of Z transforms to strengthen their understanding of signal processing; and (b) physics students and others now know and can use Z transforms, something that had not usually been part of their formal education. Based on his positive experience, Prof. Sullivan would encourage anyone using this book when teaching an FDTD course to consider this approach. However, he has left the option open to simulate dispersive methods with other techniques. The sections on Z transforms are optional and may be skipped. Appendix A on Z transforms is provided.

PROGRAMMING EXERCISES

The philosophy behind this book is that the reader will learn by doing. Therefore, most exercises involve programming. Each of Chapters 1–5 has at least one FDTD program written in Python. Each of the programs is complete and can be run as written, provided the Python interpreter and necessary libraries are installed. These programs include the graphical display of results to match many of the figures in the chapters. The programs in Chapters 1–4 are designed to be simple and procedural for ease of understanding and following the equations. Chapter 5 addresses some better Python programming practices and introduces some new features and techniques. This chapter attempts to introduce those unfamiliar with Python with some useful concepts to enhance FDTD programs and produce more readable, extendable code.

PROGRAMMING LANGUAGE

The programs in the book are written in Python. Python is a free, open-source programming language which has broad adoption in both general-purpose industries and scientific applications. This large community means that we can leverage a large number of well-documented tools and libraries. The Python libraries are constantly being expanded. Additionally, the plotting and graphical interface libraries allow the entire program to be more interactive and user-friendly, while being written in a high-level language. Libraries are also available to speed up simulations to give good performance. Python, and FDTD simulations, can be run on any modern computer.

PYTHON VERSION

All programs in this book were run with Python 3.5.1 and the following library versions:

```
matplotlib==3.0.0
numba==0.39.0
numpy==1.14.3
scipy==1.0.1
```

1

ONE-DIMENSIONAL SIMULATION WITH THE FDTD METHOD

This chapter provides a step-by-step introduction to the finite-difference time-domain (FDTD) method, beginning with the simplest possible problem, the simulation of a pulse propagating in free space in one dimension. This example is used to illustrate the FDTD formulation. Subsequent sections lead to formulations for more complicated media.

1.1 ONE-DIMENSIONAL FREE-SPACE SIMULATION

The time-dependent Maxwell's curl equations for free space are

$$\frac{\partial \boldsymbol{E}}{\partial t} = \frac{1}{\varepsilon_0} \nabla \times \boldsymbol{H}, \tag{1.1a}$$

$$\frac{\partial \boldsymbol{H}}{\partial t} = -\frac{1}{\mu_0} \nabla \times \boldsymbol{E}. \tag{1.1b}$$

Electromagnetic Simulation Using the FDTD Method with Python, Third Edition.
Jennifer E. Houle and Dennis M. Sullivan.

Published 2020 by John Wiley & Sons, Inc.

$\boldsymbol{E}$ and $\boldsymbol{H}$ are vectors in three dimensions, so, in general, Eq. (1.1a) and (1.1b) represent three equations each. We will start with a simple one-dimensional case using only E_x and H_y, so Eq. (1.1a) and (1.1b) become

$$\frac{\partial E_x}{\partial t} = -\frac{1}{\varepsilon_0}\frac{\partial H_y}{\partial z}, \tag{1.2a}$$

$$\frac{\partial H_y}{\partial t} = -\frac{1}{\mu_0}\frac{\partial E_x}{\partial z}. \tag{1.2b}$$

These are the equations of a plane wave traveling in the z direction with the electric field oriented in the x direction and the magnetic field oriented in the y direction.

Taking the central difference approximations for both the temporal and spatial derivatives gives

$$\frac{E_x^{n+1/2}(k) - E_x^{n-1/2}(k)}{\Delta t} = -\frac{1}{\varepsilon_0}\frac{H_y^n(k+\frac{1}{2}) - H_y^n(k-\frac{1}{2})}{\Delta x}, \tag{1.3a}$$

$$\frac{H_y^{n+1}(k+\frac{1}{2}) - H_y^n(k+\frac{1}{2})}{\Delta t} = -\frac{1}{\mu_0}\frac{E_x^{n+1/2}(k+1) - E_x^{n+1/2}(k)}{\Delta x}. \tag{1.3b}$$

In these two equations, time is specified by the superscripts, that is, n represents a time step, and the time t is $t = \Delta t \cdot n$. Remember, we have to discretize everything for formulation into the computer. The term $n + 1$ means one time step later. The terms in parentheses represent distance, that is, k is used to calculate the distance $z = \Delta x \cdot k$. (It might seem more sensible to use Δz as the incremental step because in this case we are going in the z direction. However, Δx is so commonly used for a spatial increment that we will use Δx.) The formulation of Eq. (1.3a) and (1.3b) assume that the E and H fields are interleaved in both space and time. H uses the arguments $k + 1/2$ and $k - 1/2$ to indicate that the H field values are assumed to be located between the E field values. This is illustrated in Fig. 1.1. Similarly, the $n + 1/2$ or $n - 1/2$ superscript indicates that it occurs slightly after or before n, respectively. Equations (1.3a) and (1.3b) can be rearranged in an iterative algorithm:

$$E_x^{n+1/2}(k) = E_x^{n-1/2}(k) - \frac{\Delta t}{\varepsilon_0 \cdot \Delta x}\left[H_y^n\left(k+\frac{1}{2}\right) - H_y^n\left(k-\frac{1}{2}\right)\right], \tag{1.4a}$$

$$H_y^{n+1}\left(k+\frac{1}{2}\right) = H_y^n\left(k+\frac{1}{2}\right) - \frac{\Delta t}{\mu_0 \cdot \Delta x}\left[E_x^{n+1/2}(k+1) - E_x^{n+1/2}(k)\right]. \tag{1.4b}$$

Notice that the calculations are interleaved in space and time. In Eq. (1.4a), for example, the new value of E_x is calculated from the previous value of E_x and the most recent values of H_y.

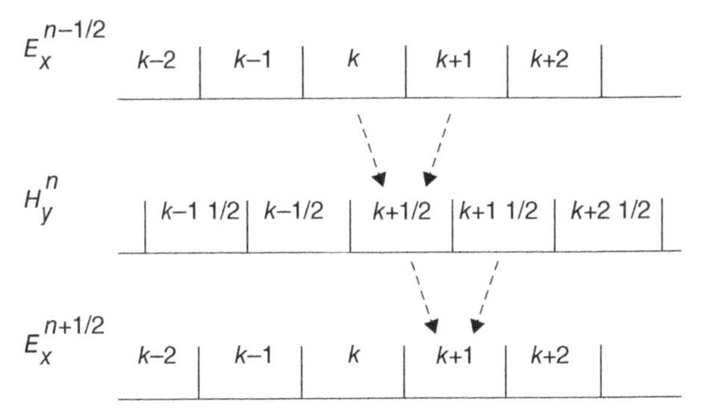

Figure 1.1 Interleaving of the E and H fields in space and time in the FDTD formulation. To calculate H_y, for instance, the neighboring values of E_x at k and $k+1$ are needed. Similarly, to calculate E_x, the values of H_y at $k+1/2$ and $k+1\frac{1}{2}$ are needed.

This is the fundamental paradigm of the FDTD method (1).

Equations (1.4a) and (1.4b) are very similar, but because ε_0 and μ_0 differ by several orders of magnitude, E_x and H_y will differ by several orders of magnitude. This is circumvented by making the following change of variables (2):

$$\widetilde{E} = \sqrt{\frac{\varepsilon_0}{\mu_0}} E. \tag{1.5}$$

Substituting this into Eq. (1.4a) and (1.4b) gives

$$\widetilde{E}_x^{n+1/2}(k) = \widetilde{E}_x^{n-1/2}(k) - \frac{\Delta t}{\sqrt{\varepsilon_0\mu_0}\cdot\Delta x}\left[H_y^n\left(k+\frac{1}{2}\right) - H_y^n\left(k-\frac{1}{2}\right)\right], \tag{1.6a}$$

$$H_y^{n+1}\left(k+\frac{1}{2}\right) = H_y^n\left(k+\frac{1}{2}\right) - \frac{\Delta t}{\sqrt{\varepsilon_0\mu_0}\cdot\Delta x}\left[\widetilde{E}_x^{n+1/2}(k+1) - \widetilde{E}_x^{n+1/2}(k)\right]. \tag{1.6b}$$

Once the cell size Δx is chosen, then the time step Δt is determined by

$$\Delta t = \frac{\Delta x}{2\cdot c_0}, \tag{1.7}$$

where c_0 is the speed of light in free space. (The reason for this will be explained in Section 1.2.) Therefore, remembering that $\varepsilon_0\mu_0 = 1/(c_0)^2$,

$$\frac{\Delta t}{\sqrt{\varepsilon_0\mu_0}\cdot\Delta x} = \frac{\Delta x}{2\cdot c_0}\cdot\frac{1}{\sqrt{\varepsilon_0\mu_0}\cdot\Delta x} = \frac{1}{2}. \tag{1.8}$$

Rewriting Eq. (1.6a) and (1.6b) in Python gives the following:

```
ex[k] = ex[k] + 0.5 * (hy[k - 1] - hy[k]),     (1.9a)

hy[k] = hy[k] + 0.5 * (ex[k] - ex[k + 1]).     (1.9b)
```

Note that the n, $n + 1/2$, or $n - 1/2$ in the superscripts is gone. Time is implicit in the FDTD method. In Eq. (1.9a), the ex on the right side of the equal sign is the previous value at $n - 1/2$, and the ex on the left side is the new value $n + 1/2$, which is being calculated. Position, however, is explicit. The only difference is that $k + 1/2$ and $k - 1/2$ are rounded to k and $k - 1$ in order to specify a position in an array in the program.

The program fd1d_1_1.py at the end of this chapter is a simple one-dimensional FDTD program. It generates a Gaussian pulse in the center of the problem space, and the pulse propagates away in both directions as seen in Fig. 1.2. The E_x field is positive in both directions, but the H_y field is negative in the negative direction. The following points are worth noting about the program:

1. The E_x and H_y values are calculated by separate loops, and they employ the interleaving described above.
2. After the E_x values are calculated, the source is calculated. This is done by simply specifying a value of E_x at the point k = kc and overriding what was previously calculated. This is referred to as a *hard source* because a specific value is imposed on the FDTD grid.

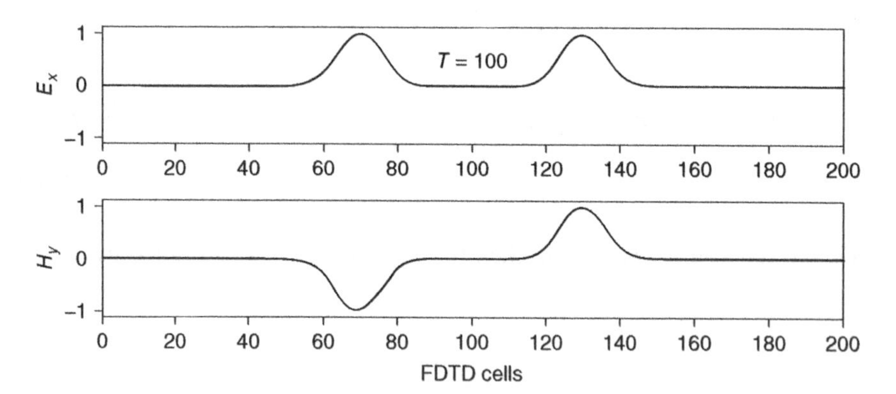

Figure 1.2 FDTD simulation of a pulse in free space after 100 time steps. The pulse originated in the center and travels outward.

PROBLEM SET 1.1

1. Get the program fd1d_1_1.py running. What happens when the pulse hits the end of the array? Why?
2. Modify the program so it has two sources, one at `kc - 20` and one at `kc + 20`. (Notice that `kc` is the center of the problem space.) What happens when the pulses meet? Explain this from basic electromagnetic (EM) theory.
3. Instead of E_x as the source, use H_y at `k = kc` as the source. What difference does it make? Try a two-point magnetic source at `kc - 1` and `kc` such that `hy[kc - 1] = - hy[kc]`. What does this look like? To what does it correspond physically?

1.2 STABILITY AND THE FDTD METHOD

Let us return to the discussion of how to determine the time step. An EM wave propagating in free space cannot go faster than the speed of light. To propagate a distance of one cell requires a minimum time of $\Delta t = \Delta x/c_0$. With a two-dimensional simulation, we must allow for the propagation in the diagonal direction, which brings the requirement to $\Delta t = \Delta x/\left(\sqrt{2}c_0\right)$. Obviously, a three-dimensional simulation requires $\Delta t = \Delta x/\left(\sqrt{3}c_0\right)$. This is summarized by the well-known *Courant Condition* (3, 4):

$$\Delta t = \frac{\Delta x}{\sqrt{n} \cdot c_0}, \tag{1.10}$$

where n is the dimension of the simulation. Unless otherwise specified, throughout this book we will determine Δt by

$$\Delta t = \frac{\Delta x}{2c_0}. \tag{1.11}$$

This is not necessarily the best formula; we will use it for simplicity to avoid using square roots.

PROBLEM SET 1.2

1. In fd1d_1_1.py, go to the governing equations, Eq. (1.9a) and (1.9b), and change the factor 0.5 to 1.0. What happens? Change it to 1.1. Now what happens? Change it to 0.25 and see what happens.

1.3 THE ABSORBING BOUNDARY CONDITION IN ONE DIMENSION

Absorbing boundary conditions are necessary to keep outgoing E and H fields from being reflected back into the problem space. Normally, in calculating the E field, we need to know the surrounding H values. This is a fundamental assumption of the FDTD method. At the edge of the problem space we will not have the value of one side. However, we have an advantage because we know that the fields at the edge must be propagating outward. We will use this fact to estimate the value at the end by using the value next to it (5).

Suppose we are looking for a boundary condition at the end where $k = 0$. If a wave is going toward a boundary in free space, it is traveling at c_0, the speed of light. So, in one time step of the FDTD algorithm, it travels

$$\text{Distance} = c_0 \cdot \Delta t = c_0 \cdot \frac{\Delta x}{2 \cdot c_0} = \frac{\Delta x}{2}.$$

This equation shows that it takes two time steps for the field to cross one cell. A commonsense approach tells us that an acceptable boundary condition might be

$$E_x^n(0) = E_x^{n-2}(1). \tag{1.12}$$

The implementation is relatively easy. Simply store a value of $E_x(1)$ for two time steps and then assign it to $E_x(0)$. Boundary conditions such as these have been implemented at both ends of the E_x array in the program fd1d_1_2.py. Figure 1.3 shows the results of a simulation using fd1d_1_2.py. A pulse that

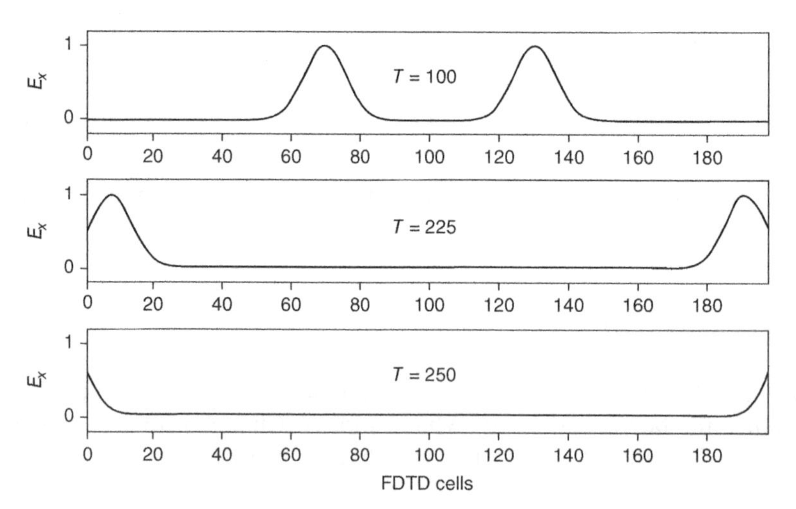

Figure 1.3 Simulation of an FDTD program with absorbing boundary conditions. Notice that the pulse is absorbed at the edges without reflecting anything back.

originates in the center propagates outward and is absorbed without reflecting anything back into the problem space.

PROBLEM SET 1.3

1. The program fd1d_1_2.py has absorbing boundary conditions at both ends. Get this program running and test it to ensure that the boundary conditions completely absorb the pulse.

1.4 PROPAGATION IN A DIELECTRIC MEDIUM

In order to simulate a medium with a dielectric constant other than 1, which corresponds to free space, we have to add the relative dielectric constant ε_r to Maxwell's equations:

$$\frac{\partial \boldsymbol{E}}{\partial t} = \frac{1}{\varepsilon_r \varepsilon_0} \nabla \times \boldsymbol{H}, \tag{1.13a}$$

$$\frac{\partial \boldsymbol{H}}{\partial t} = -\frac{1}{\mu_0} \nabla \times \boldsymbol{E}. \tag{1.13b}$$

We will stay with our one-dimensional example,

$$\frac{\partial E_x}{\partial t} = -\frac{1}{\varepsilon_r \varepsilon_0} \frac{\partial H_y}{\partial z}, \tag{1.14a}$$

$$\frac{\partial H_y}{\partial t} = -\frac{1}{\mu_0} \frac{\partial E_x}{\partial z}, \tag{1.14b}$$

then go to the finite-difference approximations and make the change of variables in Eq. (1.5):

$$\tilde{E}_x^{n+1/2}(k) = \tilde{E}_x^{n-1/2}(k) - \frac{1}{2 \cdot \varepsilon_r} \left[H_y^n\left(k + \frac{1}{2}\right) - H_y^n\left(k - \frac{1}{2}\right) \right], \tag{1.15a}$$

$$H_y^{n+1}\left(k + \frac{1}{2}\right) = H_y^n\left(k + \frac{1}{2}\right) - \frac{1}{2} \left[\tilde{E}_x^{n+1/2}(k+1) - \tilde{E}_x^{n+1/2}(k) \right]. \tag{1.15b}$$

From this we can get the computer equations

```
ex[k] = ex[k] + cb[k] * (hy[k - 1] - hy[k])          (1.16a)
hy[k] = hy[k] + 0.5 * (ex[k] - ex[k + 1]),           (1.16b)
```

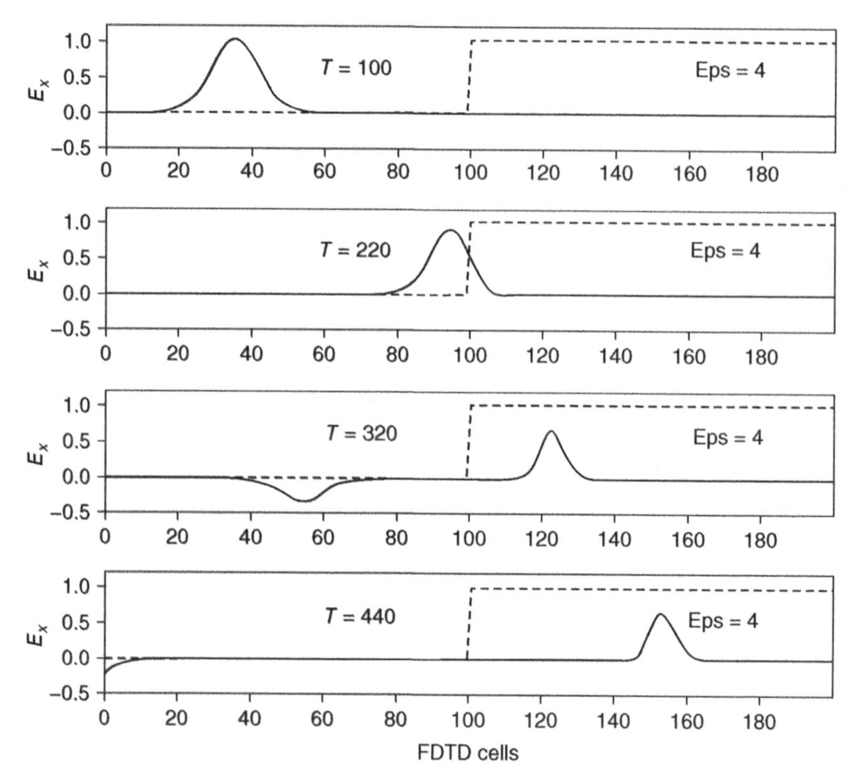

Figure 1.4 Simulation of a pulse striking dielectric material with a dielectric constant of 4. The source originates at cell number 5.

where

$$\texttt{cb[k]} = \texttt{0.5/epsilon} \tag{1.17}$$

over those values of k that specify the dielectric material.

The program fd1d_1_3.py simulates the interaction of a pulse traveling in free space until it strikes a dielectric medium. The medium is specified by the parameter `cb` in Eq. (1.17). Figure 1.4 shows the result of a simulation with a dielectric medium having a relative dielectric constant of 4. Note that one portion of the pulse propagates into the medium and the other is reflected, in keeping with basic EM theory (6).

PROBLEM SET 1.4

1. The program fd1d_1_3.py simulates a problem containing partly free space and partly dielectric material. Run this program and duplicate the results of Fig. 1.4.

2. Look at the relative amplitudes of the reflected and transmitted pulses. Are they correct? Check them by calculating the reflection and transmission coefficients (Appendix 1.A).
3. Still using a dielectric constant of 4, let the transmitted pulse propagate until it hits the far right wall. What happens? What could you do to correct this?

1.5 SIMULATING DIFFERENT SOURCES

In the fd1d_1_1.py and fd1d_1_2.py, a source is assigned as values to E_x; this is referred to as a *hard source*. In fd1d_1_3.py, however, a value is added to E_x at a certain point; this is called a *soft source*. The reason is that with a hard source, a propagating pulse will see that added value and be reflected because a hard value of E_x looks like a metal wall to FDTD. With the soft source, a propagating pulse will just pass through.

Until now, we have been using a Gaussian pulse as the source. It is very easy to switch to a sinusoidal source. Just replace the parameter `pulse` with the following:

```
pulse = sin(2 * pi * freq_in * dt * time_step)
ex[5] = pulse + ex[5].
```

The parameter `freq_in` determines the frequency of the wave. This source is used in the program fd1d_1_4.py. Figure 1.5 shows the same dielectric medium problem with a sinusoidal source. A frequency of 700 MHz is used. Notice that the simulation was stopped before the wave reached the far right side. Remember that we have an absorbing boundary condition, but only for free space.

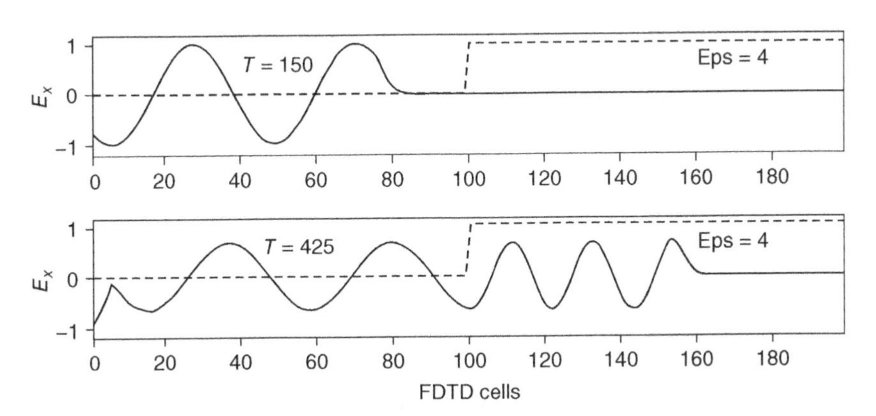

Figure 1.5 Simulation of a propagating sinusoidal wave of 700 MHz striking a medium with a relative dielectric constant of 4.

In fd1d_1_4.py, the cell size `ddx` and the time step `dt` are specified explicitly. We do this because we need `dt` in the calculation of `pulse`. The cell size `ddx` is only specified because it is needed to calculate `dt` from Eq. (1.7).

PROBLEM SET 1.5

1. Modify your program fd1d_1_3.py to simulate the sinusoidal source (see fd1d_1_4.py).
2. Keep increasing your incident frequency from 700 MHz upward at intervals of 300 MHz. What happens?
3. A *wave packet*, a sinusoidal function in a Gaussian envelope, is a type of propagating wave function that is of great interest in areas such as optics. Modify your program to simulate a wave packet.

1.6 DETERMINING CELL SIZE

Choosing the cell size to be used in an FDTD formulation is similar to any approximation procedure: Enough sampling points must be taken to ensure that an adequate representation is made. The number of points per wavelength is dependent on many factors (3, 4). However, a good rule of thumb is 10 points per wavelength. Experience has shown this to be adequate, with inaccuracies appearing as soon as the sampling drops below this rate.

Naturally, we must use a worst-case scenario. In general, this will involve looking at the highest frequencies we are simulating and determining the corresponding wavelength. For instance, suppose we are running simulations with 400 MHz. In free space, EM energy will propagate at the wavelength

$$\lambda_0 = \frac{c_0}{400\ \text{MHz}} = \frac{3 \times 10^8 \text{m/s}}{4 \times 10^8 \text{s}^{-1}} = 0.75\ \text{m}. \tag{1.18}$$

If we were only simulating free space, we would choose

$$\Delta x = \frac{\lambda_0}{10} = 7.5\,\text{cm}.$$

However, if we are simulating EM propagation in biological tissues, for instance, we must look at the wavelength in the tissue with the highest dielectric constant, because this will have the corresponding shortest wavelength. For instance, muscle has a relative dielectric constant of about 50 at 400 MHz, so

$$\lambda_0 = \frac{\left(\frac{c_0}{\sqrt{50}}\right)}{400\,\text{MHz}} = \frac{0.424 \times 10^8\,\text{m/s}}{4 \times 10^8\,\text{s}^{-1}} = 10.6\,\text{cm}.$$

In this case, we would probably select a cell size of 1 cm.

PROBLEM SET 1.6

1. Simulate a 3 GHz sine wave impinging on a material with a dielectric constant of $\varepsilon_r = 20$.

1.7 PROPAGATION IN A LOSSY DIELECTRIC MEDIUM

So far, we have simulated EM propagation in free space or in simple media that are specified by the relative dielectric constant ε_r. However, there are many media that also have a loss term specified by the conductivity. This loss term results in the attenuation of the propagating energy.

Once more we will start with the time-dependent Maxwell's curl equations, but we will write them in a more general form, which allows us to simulate propagation in media that have conductivity:

$$\varepsilon_r \varepsilon_0 \frac{\partial \boldsymbol{E}}{\partial t} = \nabla \times \boldsymbol{H} - \boldsymbol{J}, \tag{1.19a}$$

$$\frac{\partial \boldsymbol{H}}{\partial t} = -\frac{1}{\mu_0} \nabla \times \boldsymbol{E}. \tag{1.19b}$$

$\boldsymbol{J}$, the current density, can also be written as

$$\boldsymbol{J} = \sigma \boldsymbol{E},$$

where σ is the conductivity. Putting this into Eq. (1.19a) and dividing through by the dielectric constant we get

$$\frac{\partial \boldsymbol{E}}{\partial t} = \frac{1}{\varepsilon_r \varepsilon_0} \nabla \times \boldsymbol{H} - \frac{\sigma}{\varepsilon_r \varepsilon_0} \boldsymbol{E}.$$

We now revert to our simple one-dimensional equation:

$$\frac{\partial E_x(t)}{\partial t} = -\frac{1}{\varepsilon_r \varepsilon_0} \frac{\partial H_y(t)}{\partial z} - \frac{\sigma}{\varepsilon_r \varepsilon_0} E_x(t),$$

and make the change of variable in Eq. (1.5), which gives

$$\frac{\partial \widetilde{E}_x(t)}{\partial t} = -\frac{1}{\varepsilon_r\sqrt{\mu_0\varepsilon_0}}\frac{\partial H_y(t)}{\partial z} - \frac{\sigma}{\varepsilon_r\varepsilon_0}\widetilde{E}_x(t), \tag{1.20a}$$

$$\frac{\partial H_y(t)}{\partial t} = -\frac{1}{\sqrt{\mu_0\varepsilon_0}}\frac{\partial \widetilde{E}_x(t)}{\partial z}. \tag{1.20b}$$

Next, take the finite-difference approximation for both the temporal and spatial derivatives similar to Eq. (1.3a):

$$\begin{aligned}\frac{E_x^{n+1/2}(k) - E_x^{n-1/2}(k)}{\Delta t} = &-\frac{1}{\varepsilon_r\sqrt{\varepsilon_0\mu_0}}\frac{H_y^n\left(k+\frac{1}{2}\right) - H_y^n\left(k-\frac{1}{2}\right)}{\Delta x} \\ &-\frac{\sigma}{\varepsilon_r\varepsilon_0}\frac{E_x^{n+1/2}(k) + E_x^{n-1/2}(k)}{2}.\end{aligned} \tag{1.21}$$

Notice that the last term in Eq. (1.20a) is approximated as the average across two time steps in Eq. (1.21). The tildes were dropped from Eq. (1.21) for simplicity. From Eq. (1.8),

$$\frac{1}{\sqrt{\varepsilon_0\mu_0}}\frac{\Delta t}{\Delta x} = \frac{1}{2},$$

so Eq. (1.21) becomes

$$\begin{aligned}E_x^{n+1/2}(k)\left(1 + \frac{\Delta t\cdot\sigma}{2\varepsilon_r\varepsilon_0}\right) = &\left(1 - \frac{\Delta t\cdot\sigma}{2\varepsilon_r\varepsilon_0}\right)E_x^{n-1/2}(k) \\ &-\frac{\left(\frac{1}{2}\right)}{\varepsilon_r}\left[H_y^n\left(k+\frac{1}{2}\right) - H_y^n\left(k-\frac{1}{2}\right)\right]\end{aligned}$$

or

$$\begin{aligned}E_x^{n+1/2}(k) = &\frac{\left(1 - \frac{\Delta t\cdot\sigma}{2\varepsilon_r\varepsilon_0}\right)}{\left(1 + \frac{\Delta t\cdot\sigma}{2\varepsilon_r\varepsilon_0}\right)}E_x^{n-1/2}(k) \\ &-\frac{\left(\frac{1}{2}\right)}{\varepsilon_r\left(1 + \frac{\Delta t\cdot\sigma}{2\varepsilon_r\varepsilon_0}\right)}\left[H_y^n\left(k+\frac{1}{2}\right) - H_y^n\left(k-\frac{1}{2}\right)\right].\end{aligned}$$

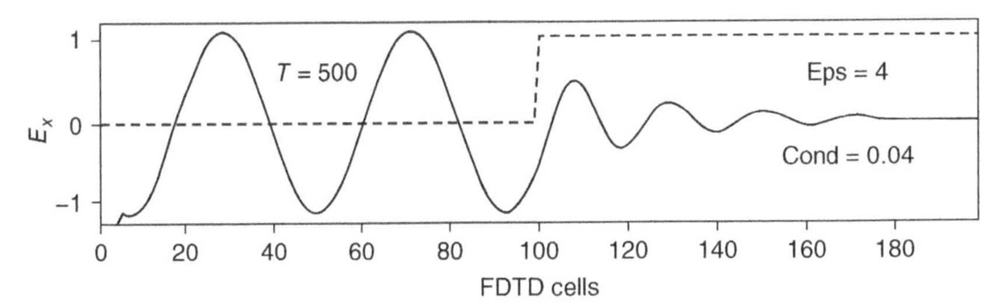

Figure 1.6 Simulation of a propagating sinusoidal wave striking a lossy dielectric material with a dielectric constant of 4 and a conductivity of 0.04 (S/m). The source is 700 MHz and originates at cell number 5.

From these we can get the computer equations:

```
ex[k] = ca[k] * ex[k] + cb[k] * (hy[k - 1] - hy[k])        (1.22a)
hy[k] = hy[k] + 0.5 * (ex[k] - ex[k + 1]),                   (1.22b)
```

where

```
  eaf = dt * sigma/(2 * epsz * epsilon),                     (1.23a)
ca[k] = (1 - eaf)/(1 + eaf),                                 (1.23b)
cb[k] = 0.5/(epsilon * (1 + eaf)).                           (1.23c)
```

The program fd1d_1_5.py simulates a sinusoidal wave hitting a lossy medium that has a dielectric constant of 4 and a conductivity of 0.04. The pulse is generated at the left side and propagates to the right (Fig. 1.6). Notice that the waveform in the medium is absorbed before it hits the boundary, so we do not have to worry about absorbing boundary conditions.

PROBLEM SET 1.7

1. Run program fd1d_1_5.py to simulate a complex dielectric material. Duplicate the results of Fig. 1.6.
2. Verify that your calculation of the sine wave in the lossy dielectric is correct: That is, it is the correct amplitude going into the slab, and then it attenuates at the proper rate (Appendix 1.A).
3. How would you write an absorbing boundary condition for a lossy material?
4. Simulate a pulse hitting a metal wall. This is very easy to do, if you remember that metal has a very high conductivity. For the complex dielectric, just

use $\sigma = 1e6$ or any large number. (It does not have to be the correct conductivity of the metal, just very large.) What does this do to the FDTD parameters `ca` and `cb`? What result does this have for the field parameters E_x and H_y? If you did not want to specify dielectric parameters, how else would you simulate metal in an FDTD program?

1.A APPENDIX

When a plane wave traveling in medium 1 strikes medium 2, the fraction that is reflected is given by the reflection coefficient Γ, and the fraction that is transmitted into medium 2 is given by the transmission coefficient τ. These are determined by the intrinsic impedances η_1 and η_2 of the respective media (6):

$$\Gamma = \frac{E_{\mathrm{ref}}}{E_{\mathrm{inc}}} = \frac{\eta_2 - \eta_1}{\eta_2 + \eta_1} \tag{1.A.1}$$

$$\tau = \frac{E_{\mathrm{trans}}}{E_{\mathrm{inc}}} = \frac{2\eta_2}{\eta_2 + \eta_1}. \tag{1.A.2}$$

The impedances are given by

$$\eta = \sqrt{\frac{\mu}{\varepsilon_0 \varepsilon_r}}. \tag{1.A.3}$$

The complex relative dielectric constant ε_r^* is given by

$$\varepsilon_r^* = \varepsilon_r + \frac{\sigma}{j\omega\varepsilon_0}.$$

For the case where $\mu = \mu_0$, Eq. (1.A.1) and Eq. (1.A.2) become

$$\Gamma = \frac{\frac{1}{\sqrt{\varepsilon_2^*}} - \frac{1}{\sqrt{\varepsilon_1^*}}}{\frac{1}{\sqrt{\varepsilon_2^*}} + \frac{1}{\sqrt{\varepsilon_1^*}}} = \frac{\sqrt{\varepsilon_1^*} - \sqrt{\varepsilon_2^*}}{\sqrt{\varepsilon_1^*} + \sqrt{\varepsilon_2^*}} \tag{1.A.4}$$

$$\tau = \frac{\frac{2}{\sqrt{\varepsilon_1^*}}}{\frac{1}{\sqrt{\varepsilon_2^*}} + \frac{1}{\sqrt{\varepsilon_1^*}}} = \frac{2\sqrt{\varepsilon_1^*}}{\sqrt{\varepsilon_1^*} + \sqrt{\varepsilon_2^*}}. \tag{1.A.5}$$

The amplitude of an electric field propagating in the positive z direction in a lossy dielectric medium is given by

$$E_x(z) = E_0 \cdot \mathrm{e}^{-k \cdot z} = E_0 \cdot \mathrm{e}^{-\,\mathrm{Re}\{k\} \cdot z}\mathrm{e}^{-\,j\mathrm{Im}\{k\} \cdot z},$$

where E_0 is the amplitude at $z = 0$. The wave number k is determined by

$$k = \frac{\omega}{c_0}\sqrt{\varepsilon_r}. \tag{1.A.6}$$

REFERENCES

1. K. S. Yee, Numerical solution of initial boundary value problems involving Maxwell's equations in isotropic media, *IEEE Trans. Antennas Propag.*, vol. 17, 1966, pp. 585–589.
2. A. Taflove and M. Brodwin, Numerical solution of steady state electromagnetic scattering problems using the time-dependent Maxwell's equations, *IEEE Trans. Microwave Theory Tech.*, vol. 23, 1975, pp. 623–730.
3. A. Taflove, *Computational Electrodynamics: The Finite-Difference Time-Domain Method*, 3rd Edition, Boston, MA: Artech House, 1995.
4. K. S. Kunz and R. J. Luebbers, *The Finite Difference Time Domain Method for Electromagnetics*, Boca Raton, FL: CRC Press, 1993.
5. G. Mur, Absorbing boundary conditions for the finite-difference approximation of the time domain electromagnetic field equations, *IEEE Trans. Electromagn. Compat.*, vol. 23, 1981, pp. 377–384.
6. D. K. Cheng, *Field and Wave Electromagnetics, Menlo Park, CA: Addison-Wesley, 1992.*

PYTHON PROGRAMS USED TO GENERATE FIGURES IN THIS CHAPTER

```
""" fd3d_1_1.py: 1D FDTD

Simulation in free space
"""

import numpy as np
from math import exp
from matplotlib import pyplot as plt

ke = 200
ex = np.zeros(ke)
hy = np.zeros(ke)

# Pulse parameters
kc = int(ke / 2)
t0 = 40
spread = 12
```

```
nsteps = 100

# Main FDTD Loop
for time_step in range(1, nsteps + 1):

    # Calculate the Ex field
    for k in range(1, ke):
        ex[k] = ex[k] + 0.5 * (hy[k - 1] - hy[k])

    # Put a Gaussian pulse in the middle
    pulse = exp(-0.5 * ((t0 - time_step) / spread) ** 2)
    ex[kc] = pulse

    # Calculate the Hy field
    for k in range(ke - 1):
        hy[k] = hy[k] + 0.5 * (ex[k] - ex[k + 1])

# Plot the outputs as shown in Fig. 1.2
plt.rcParams['font.size'] = 12
plt.figure(figsize=(8, 3.5))

plt.subplot(211)
plt.plot(ex, color='k', linewidth=1)
plt.ylabel('E$_x$', fontsize='14')
plt.xticks(np.arange(0, 201, step=20))
plt.xlim(0, 200)
plt.yticks(np.arange(-1, 1.2, step=1))
plt.ylim(-1.2, 1.2)
plt.text(100, 0.5, 'T = {}'.format(time_step),
horizontalalignment='center')

plt.subplot(212)
plt.plot(hy, color='k', linewidth=1)
plt.ylabel('H$_y$', fontsize='14')
plt.xlabel('FDTD cells')
plt.xticks(np.arange(0, 201, step=20))
plt.xlim(0, 200)
plt.yticks(np.arange(-1, 1.2, step=1))
plt.ylim(-1.2, 1.2)

plt.subplots_adjust(bottom=0.2, hspace=0.45)
plt.show()


""" fd3d_1_2.py: 1D FDTD

Simulation in free space
Absorbing Boundary Condition added
"""
```

```
import numpy as np
from math import exp
from matplotlib import pyplot as plt

ke = 200
ex = np.zeros(ke)
hy = np.zeros(ke)

# Pulse parameters
kc = int(ke / 2)
t0 = 40
spread = 12

boundary_low = [0, 0]
boundary_high = [0, 0]

nsteps = 250

# Dictionary to keep track of desired points for plotting
plotting_points = [
    {'num_steps': 100, 'data_to_plot': None, 'label': ''},
    {'num_steps': 225, 'data_to_plot': None, 'label': ''},
    {'num_steps': 250, 'data_to_plot': None, 'label': 'FDTD cells'}
]

# Main FDTD Loop
for time_step in range(1, nsteps + 1):

    # Calculate the Ex field
    for k in range(1, ke):
        ex[k] = ex[k] + 0.5 * (hy[k - 1] - hy[k])

    # Put a Gaussian pulse in the middle
    pulse = exp(-0.5 * ((t0 - time_step) / spread) ** 2)
    ex[kc] = pulse

    # Absorbing Boundary Conditions
    ex[0] = boundary_low.pop(0)
    boundary_low.append(ex[1])

    ex[ke - 1] = boundary_high.pop(0)
    boundary_high.append(ex[ke - 2])

    # Calculate the Hy field
    for k in range(ke - 1):
        hy[k] = hy[k] + 0.5 * (ex[k] - ex[k + 1])
```

```
    # Save data at certain points for later plotting
    for plotting_point in plotting_points:
        if time_step == plotting_point['num_steps']:
            plotting_point['data_to_plot'] = np.copy(ex)

# Plot the outputs as shown in Fig. 1.3
plt.rcParams['font.size'] = 12
fig = plt.figure(figsize=(8, 5.25))


def plot_e_field(data, timestep, label):
    """Plot of E field at a single time step"""

    plt.plot(data, color='k', linewidth=1)
    plt.ylabel('E$_x$', fontsize='14')
    plt.xticks(np.arange(0, 199, step=20))
    plt.xlim(0, 199)
    plt.yticks(np.arange(0, 1.2, step=1))
    plt.ylim(-0.2, 1.2)
    plt.text(100, 0.5, 'T = {}'.format(timestep),
             horizontalalignment='center')
    plt.xlabel('{}'.format(label))


# Plot the E field at each of the time steps saved earlier
for subplot_num, plotting_point in enumerate(plotting_points):
    ax = fig.add_subplot(3, 1, subplot_num + 1)
    plot_e_field(plotting_point['data_to_plot'],
                 plotting_point['num_steps'],
                 plotting_point['label'])

plt.tight_layout()
plt.show()
```

```
""" fd3d_1_3.py: 1D FDTD

Simulation of a pulse hitting a dielectric medium
"""

import numpy as np
from math import exp
from matplotlib import pyplot as plt

ke = 200
ex = np.zeros(ke)
hy = np.zeros(ke)

t0 = 40
spread = 12
```

```
boundary_low = [0, 0]
boundary_high = [0, 0]

# Create Dielectric Profile
cb = np.ones(ke)
cb = 0.5 * cb
cb_start = 100
epsilon = 4
cb[cb_start:] = 0.5 / epsilon

nsteps = 440

# Dictionary to keep track of desired points for plotting
plotting_points = [
    {'num_steps': 100, 'data_to_plot': None, 'label': ''},
    {'num_steps': 220, 'data_to_plot': None, 'label': ''},
    {'num_steps': 320, 'data_to_plot': None, 'label': ''},
    {'num_steps': 440, 'data_to_plot': None, 'label': 'FDTD cells'}
]

# Main FDTD Loop
for time_step in range(1, nsteps + 1):

    # Calculate the Ex field
    for k in range(1, ke):
        ex[k] = ex[k] + cb[k] * (hy[k - 1] - hy[k])

    # Put a Gaussian pulse at the low end
    pulse = exp(-0.5 * ((t0 - time_step) / spread) ** 2)
    ex[5] = pulse + ex[5]

    # Absorbing Boundary Conditions
    ex[0] = boundary_low.pop(0)
    boundary_low.append(ex[1])

    ex[ke - 1] = boundary_high.pop(0)
    boundary_high.append(ex[ke - 2])

    # Calculate the Hy field
    for k in range(ke - 1):
        hy[k] = hy[k] + 0.5 * (ex[k] - ex[k + 1])

    # Save data at certain points for later plotting
    for plotting_point in plotting_points:
        if time_step == plotting_point['num_steps']:
            plotting_point['data_to_plot'] = np.copy(ex)
```

```
# Plot the outputs as shown in Fig. 1.4
plt.rcParams['font.size'] = 12
fig = plt.figure(figsize=(8, 7))


def plot_e_field(data, timestep, epsilon, cb, label):
    """Plot of E field at a single time step"""

    plt.plot(data, color='k', linewidth=1)
    plt.ylabel('E$_x$', fontsize='14')
    plt.xticks(np.arange(0, 199, step=20))
    plt.xlim(0, 199)
    plt.yticks(np.arange(-0.5, 1.2, step=0.5))
    plt.ylim(-0.5, 1.2)
    plt.text(70, 0.5, 'T = {}'.format(timestep),
             horizontalalignment='center')
    plt.plot((0.5 / cb - 1) / 3, 'k--', linewidth=0.75)
    # The math on cb above is just for scaling
    plt.text(170, 0.5, 'Eps = {}'.format(epsilon),
             horizontalalignment='center')
    plt.xlabel('{}'.format(label))


# Plot the E field at each of the time steps saved earlier
for subplot_num, plotting_point in enumerate(plotting_points):
    ax = fig.add_subplot(4, 1, subplot_num + 1)
    plot_e_field(plotting_point['data_to_plot'],
                 plotting_point['num_steps'], epsilon, cb,
                 plotting_point['label'])

plt.subplots_adjust(bottom=0.1, hspace=0.45)
plt.show()


""" fd3d_1_4.py: 1D FDTD

Simulation of a sinusoidal wave hitting a dielectric medium
"""

import numpy as np
from math import pi, sin
from matplotlib import pyplot as plt

ke = 200
ex = np.zeros(ke)
hy = np.zeros(ke)
```

```
ddx = 0.01  # Cell size
dt = ddx / 6e8  # Time step size
freq_in = 700e6

boundary_low = [0, 0]
boundary_high = [0, 0]

# Create Dielectric Profile
cb = np.ones(ke)
cb = 0.5 * cb
cb_start = 100
epsilon = 4
cb[cb_start:] = 0.5 / epsilon

nsteps = 425

# Dictionary to keep track of desired points for plotting
plotting_points = [
    {'num_steps': 150, 'data_to_plot': None, 'label': ''},
    {'num_steps': 425, 'data_to_plot': None, 'label': 'FDTD cells'}
]

# Main FDTD Loop
for time_step in range(1, nsteps + 1):

    # Calculate the Ex field
    for k in range(1, ke):
        ex[k] = ex[k] + cb[k] * (hy[k - 1] - hy[k])

    # Put a sinusoidal at the low end
    pulse = sin(2 * pi * freq_in * dt * time_step)
    ex[5] = pulse + ex[5]

    # Absorbing Boundary Conditions
    ex[0] = boundary_low.pop(0)
    boundary_low.append(ex[1])

    ex[ke - 1] = boundary_high.pop(0)
    boundary_high.append(ex[ke - 2])

    # Calculate the Hy field
    for k in range(ke - 1):
        hy[k] = hy[k] + 0.5 * (ex[k] - ex[k + 1])

    # Save data at certain points for later plotting
    for plotting_point in plotting_points:
        if time_step == plotting_point['num_steps']:
            plotting_point['data_to_plot'] = np.copy(ex)
```

```
# Plot the outputs in Fig. 1.5
plt.rcParams['font.size'] = 12
fig = plt.figure(figsize=(8, 3.5))


def plot_e_field(data, timestep, epsilon, cb, label):
    """Plot of E field at a single time step"""

    plt.plot(data, color='k', linewidth=1)
    plt.ylabel('E$_x$', fontsize='14')
    plt.xticks(np.arange(0, 199, step=20))
    plt.xlim(0, 199)
    plt.yticks(np.arange(-1, 1.2, step=1))
    plt.ylim(-1.2, 1.2)
    plt.text(50, 0.5, 'T = {}'.format(timestep),
             horizontalalignment='center')
    plt.plot((0.5 / cb - 1) / 3, 'k--', linewidth=0.75)
    # The math on cb above is just for scaling
    plt.text(170, 0.5, 'Eps = {}'.format(epsilon),
             horizontalalignment='center')
    plt.xlabel('{}'.format(label))


# Plot the E field at each of the time steps saved earlier
for subplot_num, plotting_point in enumerate(plotting_points):
    ax = fig.add_subplot(2, 1, subplot_num + 1)
    plot_e_field(plotting_point['data_to_plot'],
                 plotting_point['num_steps'], epsilon, cb,
                 plotting_point['label'])

plt.subplots_adjust(bottom=0.2, hspace=0.45)
plt.show()


""" fd3d_1_5.py: 1D FDTD

Simulation of a sinusoid wave hitting a lossy dielectric
"""

import numpy as np
from math import pi, sin
from matplotlib import pyplot as plt

ke = 200
ex = np.zeros(ke)
hy = np.zeros(ke)
```

```
ddx = 0.01  # Cell size
dt = ddx / 6e8  # Time step size
freq_in = 700e6

boundary_low = [0, 0]
boundary_high = [0, 0]

# Create Dielectric Profile
epsz = 8.854e-12
epsilon = 4
sigma = 0.04

ca = np.ones(ke)
cb = np.ones(ke) * 0.5
cb_start = 100

eaf = dt * sigma / (2 * epsz * epsilon)
ca[cb_start:] = (1 - eaf) / (1 + eaf)
cb[cb_start:] = 0.5 / (epsilon * (1 + eaf))

nsteps = 500

# Main FDTD Loop
for time_step in range(1, nsteps + 1):

    # Calculate the Ex field
    for k in range(1, ke):
        ex[k] = ca[k] * ex[k] + cb[k] * (hy[k - 1] - hy[k])

    # Put a sinusoidal at the low end
    pulse = sin(2 * pi * freq_in * dt * time_step)
    ex[5] = pulse + ex[5]

    # Absorbing Boundary Conditions
    ex[0] = boundary_low.pop(0)
    boundary_low.append(ex[1])

    ex[ke - 1] = boundary_high.pop(0)
    boundary_high.append(ex[ke - 2])

    # Calculate the Hy field
    for k in range(ke - 1):
        hy[k] = hy[k] + 0.5 * (ex[k] - ex[k + 1])

# Plot the outputs in Fig. 1.6
plt.rcParams['font.size'] = 12
plt.figure(figsize=(8, 2.25))
```

```
plt.plot(ex, color='k', linewidth=1)
plt.ylabel('E$_x$', fontsize='14')
plt.xticks(np.arange(0, 199, step=20))
plt.xlim(0, 199)
plt.yticks(np.arange(-1, 1.2, step=1))
plt.ylim(-1.2, 1.2)
plt.text(50, 0.5, 'T = {}'.format(time_step),
         horizontalalignment='center')
plt.plot((0.5 / cb - 1) / 3, 'k--',
         linewidth=0.75)  # The math on cb is just for scaling
plt.text(170, 0.5, 'Eps = {}'.format(epsilon),
         horizontalalignment='center')
plt.text(170, -0.5, 'Cond = {}'.format(sigma),
         horizontalalignment='center')
plt.xlabel('FDTD cells')

plt.subplots_adjust(bottom=0.25, hspace=0.45)
plt.show()
```

2

MORE ON ONE-DIMENSIONAL SIMULATION

Before moving on to two- and three-dimensional problems, we will stay with one-dimensional simulation to introduce some advanced concepts. First, we will change the formulation slightly and introduce the use of the flux density into the simulation. This may initially seem like an unnecessary complication. However, as we get to frequency-dependent materials in Section 2.3, the advantages will become apparent. Then, in Section 2.2, we introduce the use of the discrete Fourier transform in finite-difference time-domain (FDTD) simulation. This is an extremely powerful method to quantify the output of the simulation.

This chapter should make apparent how closely signal processing is linked to time-domain EM simulation, especially in Section 2.4, where we use Z transforms to simulate complicated media.

2.1 REFORMULATION USING THE FLUX DENSITY

Until now, we have been using the form of Maxwell's equations given in Eq. (1.1), which uses only the $\boldsymbol{E}$ and $\boldsymbol{H}$ fields. However, a more general form is

$$\frac{\partial \boldsymbol{D}}{\partial t} = \nabla \times \boldsymbol{H}, \tag{2.1a}$$

Electromagnetic Simulation Using the FDTD Method with Python, Third Edition.
Jennifer E. Houle and Dennis M. Sullivan.

Published 2020 by John Wiley & Sons, Inc.

$$D(\omega) = \varepsilon_0 \cdot \varepsilon_r^*(\omega) \cdot E(\omega), \tag{2.1b}$$

$$\frac{\partial H}{\partial t} = -\frac{1}{\mu_0} \nabla \times E, \tag{2.1c}$$

where D is the electric flux density. Notice that Eq. (2.1b) is written in the frequency domain. The reason for this will be explained in Section 2.3. We will begin by normalizing these equations, using

$$\widetilde{E} = \sqrt{\frac{\varepsilon_0}{\mu_0}} \cdot E, \tag{2.2a}$$

$$\widetilde{D} = \sqrt{\frac{1}{\varepsilon_0 \mu_0}} \cdot D, \tag{2.2b}$$

which leads to

$$\frac{\partial \widetilde{D}}{\partial t} = \frac{1}{\sqrt{\varepsilon_0 \mu_0}} \nabla \times H, \tag{2.3a}$$

$$\widetilde{D}(\omega) = \varepsilon_r^*(\omega) \cdot \widetilde{E}(\omega), \tag{2.3b}$$

$$\frac{\partial H}{\partial t} = -\frac{1}{\sqrt{\varepsilon_0 \mu_0}} \nabla \times \widetilde{E}. \tag{2.3c}$$

We saw in Chapter 1 that this form of Eq. (2.3a) and (2.3c) leads to the very simple finite-difference equations Eq. (1.4a) and (1.4b). The only change is the use of D instead of E. However, we still have to get Eq. (2.3b) into a time-domain difference equation for implementation into FDTD. The first task is to shift from the frequency domain to the time domain. We will assume we are dealing with a lossy dielectric medium of the form

$$\varepsilon_r^*(\omega) = \varepsilon_r + \frac{\sigma}{j\omega \cdot \varepsilon_0} \tag{2.4}$$

and substitute Eq. (2.4) into Eq. (2.3b):

$$D(\omega) = \varepsilon_r E(\omega) + \frac{\sigma}{j\omega \cdot \varepsilon_0} E(\omega). \tag{2.5}$$

Notice the tildes were removed, but E and D for the rest of this chapter refer to the modified values. Taking the first term into the time domain is not a problem because it is a simple multiplication. In the second term, Fourier theory tells us that $1/(j\omega)$ in the frequency domain is integration in the time domain, so Eq. (2.5) becomes

$$\boldsymbol{D}(t) = \varepsilon_r \boldsymbol{E}(t) + \frac{\sigma}{\varepsilon_0} \int_0^t \boldsymbol{E}(t')\ dt'.$$

We will want to go to the sampled time domain, so the integral will be approximated as a summation over the time steps Δt:

$$\boldsymbol{D}^n = \varepsilon_r \boldsymbol{E}^n + \frac{\sigma \cdot \Delta t}{\varepsilon_0} \sum_{i=0}^{n} \boldsymbol{E}^i. \tag{2.6}$$

Note that $\boldsymbol{E}$ and $\boldsymbol{D}$ are specified at time $t = n \cdot \Delta t$. There is one problem: Looking back at Eq. (2.3b), we see that we will have to solve for $\boldsymbol{E}^n$ given the value $\boldsymbol{D}^n$. But the value $\boldsymbol{E}^n$ is needed in the calculation of the summation. We will circumvent this by separating the $\boldsymbol{E}^n$ term from the rest of the summation:

$$\boldsymbol{D}^n = \varepsilon_r \boldsymbol{E}^n + \frac{\sigma \cdot \Delta t}{\varepsilon_0} \boldsymbol{E}^n + \frac{\sigma \cdot \Delta t}{\varepsilon_0} \sum_{i=0}^{n-1} \boldsymbol{E}^i.$$

Now, we can calculate $\boldsymbol{E}^n$ from

$$\boldsymbol{E}^n = \frac{\boldsymbol{D}^n - \frac{\sigma \cdot \Delta t}{\varepsilon_0} \sum_{i=0}^{n-1} \boldsymbol{E}^i}{\varepsilon_r + \frac{\sigma \cdot \Delta t}{\varepsilon_0}}. \tag{2.7}$$

We can calculate $\boldsymbol{E}^n$, the *current* value of $\boldsymbol{E}$, from the current value of $\boldsymbol{D}$ and *previous* values of $\boldsymbol{E}$. It will prove advantageous to define a new parameter for the summation

$$\boldsymbol{I}^{n-1} = \frac{\sigma \cdot \Delta t}{\varepsilon_0} \sum_{i=0}^{n-1} \boldsymbol{E}^i,$$

so Eq. (2.7) can be reformulated with the following two equations:

$$\boldsymbol{E}^n = \frac{\boldsymbol{D}^n - \boldsymbol{I}^{n-1}}{\varepsilon_r + \frac{\sigma \cdot \Delta t}{\varepsilon_0}} \tag{2.8a}$$

$$\boldsymbol{I}^n = \boldsymbol{I}^{n-1} + \frac{\sigma \cdot \Delta t}{\varepsilon_0} \boldsymbol{E}^n. \tag{2.8b}$$

Note that the summation is calculated by Eq. (2.8b), which at every time step n simply adds the value $\boldsymbol{E}^n$ times the constant term to the previous values of the summation at $n-1$. It is not necessary to store all the values of $\boldsymbol{E}^n$ from 0 to n. Now the entire FDTD formulation in one dimension, using the same orientation that was used in Chapter 1, is

```
dx[k] = dx[k] + 0.5 * (hy[k - 1] - hy[k])      (2.9a)

ex[k] = gax[k] * (dx[k] - ix[k])               (2.9b)

ix[k] = ix[k] + gbx[k] * ex[k]                 (2.9c)

hy[k] = hy[k] + 0.5 * (ex[k] - ex[k + 1])      (2.9d)
```

where

```
gax[k] = 1/(epsr + (sigma * dt/epsz))          (2.10a)

gbx[k] = sigma * dt/epsz .                     (2.10b)
```

The important point is this: all of the information regarding the media is contained in Eq. (2.9b) and (2.9c). For free space, `gax = 1` and `gbx = 0`; for lossy material, `gax` and `gbx` are calculated according to Eq. (2.10a) and (2.10b). In calculating `ex [k]` at the point k, we use only values of `dx [k]` and the previous values of `ex [k]` in the time domain. Equation (2.9a) and (2.9d), which contain the spatial derivatives, do not change regardless of the media.

It may seem as though we have paid a high price for this fancy formulation compared to the formulation of Chapter 1. We now need D_x as well as E_x and an auxiliary parameter I_x. The real advantage comes when we deal with more complicated materials, as we will see in the following sections.

PROBLEM SET 2.1

1. The program fd1d_2_1.py implements the reformulation using the flux density. Get this program running and repeat the results of Problem 1.7.1.

2.2 CALCULATING THE FREQUENCY DOMAIN OUTPUT

Until now, the output of our FDTD programs has been the $\boldsymbol{E}$ field itself, and we have been content to simply watch a pulse or sine wave propagate through various media. Needless to say, before any such practical applications can be implemented, it will be necessary to quantify the results. Suppose now that we are asked to calculate the ***E*** field distribution at every point in a dielectric medium subject to illumination at various frequencies. One approach would be to use a sinusoidal source and iterate the FDTD program until a steady state is reached and determine the resulting amplitude and phase at every point of interest in the medium. This would work, but then we must repeat the process for every frequency of interest. According to system theory, we can get the response to every frequency if we use an impulse as the source.

We could go back to the Gaussian pulse, which, if narrow enough, is a good approximation of an impulse. We then iterate the FDTD program until the pulse has died out, and take the Fourier transform of the ***E*** field in the medium. If we have the Fourier transform of the ***E*** field at a point, then we know the amplitude and phase of the ***E*** field that would result from illumination by any sinusoidal source. This, too, has a serious drawback: the ***E*** field for all the time domain data at every point of interest would have to be stored until the FDTD program is through iterating so the Fourier transform of the data could be taken, presumably using a fast Fourier transform algorithm. This presents a logistical difficulty.

Here is an alternative. Suppose we want to calculate the Fourier transform of the ***E*** field $\boldsymbol{E}(t)$ at a frequency f_1. This can be done by the equation

$$\boldsymbol{E}(f_1) = \int_0^{t_T} \boldsymbol{E}(t) \cdot \mathrm{e}^{-j2\pi \cdot f_1 \cdot t} \mathrm{d}t. \tag{2.11}$$

Notice that the lower limit of the integral is at 0 because the FDTD program assumes all causal functions. The upper limit is t_T, the time at which the FDTD iteration is halted. Rewriting Eq. (2.11) in a finite-difference form,

$$\boldsymbol{E}(f_1) = \sum_{n=0}^{T} \boldsymbol{E}(n \cdot \Delta t) \cdot \mathrm{e}^{-j2\pi \cdot f_1 (n \cdot \Delta t)}, \tag{2.12}$$

where T is the number of iterations and Δt is the time step, so $t_T = \Delta t \cdot T$. Equation (2.12) can be divided into its real and imaginary parts

$$\begin{aligned} \boldsymbol{E}(f_1) = & \sum_{n=0}^{T} \boldsymbol{E}(n \cdot \Delta t) \cdot \cos(2\pi f_1 \cdot \Delta t \cdot n) \\ & - j \sum_{n=0}^{T} \boldsymbol{E}(n \cdot \Delta t) \cdot \sin(2\pi f_1 \cdot \Delta t \cdot n), \end{aligned} \tag{2.13}$$

which can be implemented in computer code by

$$\texttt{real\_pt[m,k]} = \texttt{real\_pt[m,k]} + \texttt{cos (2 * pi * freq[m]} \\ \texttt{* dt * time\_step) * ex[k]} \tag{2.14a}$$

$$\texttt{imag\_pt[m,k]} = \texttt{imag\_pt[m,k]} - \texttt{sin (2 * pi * freq[m]} \\ \texttt{* dt * time\_step) * ex[k]} . \tag{2.14b}$$

For every point k in the region of interest, we require only two computer words of memory for every frequency of interest f_m. At any point k, from the real part of $\boldsymbol{E}(f_m)$, `real_pt [m, k]`, and the imaginary part, `imag_pt [m, k]`, we can determine the amplitude and phase at the frequency f_m:

$$\texttt{amp[m,k]} = \texttt{sqrt((real\_pt[m,k]) * * 2 + (imag\_pt[m,k]) * * 2)} \tag{2.15a}$$

$$\texttt{phase[m,k]} = \texttt{atan2(imag\_pt[m,k], real\_pt[m,k])} . \tag{2.15b}$$

Note that there is an amplitude and phase associated with every frequency at each cell (1, 2). The program fd1d_2_2.py following the references in this chapter calculates the frequency response at three frequencies throughout the problem space. Figure 2.1 is a simulation of a pulse hitting a dielectric medium with a dielectric constant of 4, similar to Fig. 1.4. The frequency response at 500 MHz is also shown. At $T = 200$, before the pulse has hit the medium, the frequency response is one through that part of the space where the pulse has traveled. After 400 time steps, the pulse has hit the medium; some of the pulse has penetrated into the medium and some of it has been reflected. The amplitude of the transmitted pulse is determined by Eq. (1.A.5)

$$\tau = \frac{2 \cdot \sqrt{1}}{\sqrt{1} + \sqrt{4}} = 0.667,$$

which is the Fourier amplitude in the medium. The Fourier amplitude outside the medium varies between $1 - 0.333$ and $1 + 0.333$. This is in keeping with the pattern formed by the standing wave that is created from a sinusoidal signal, the reflected wave of which interacts with the original incident wave.

PROBLEM SET 2.2

1. The program fd1d_2_2.py implements the discrete Fourier transform with a Gaussian pulse as its source. Get this program running. Duplicate the results in Fig. 2.1.

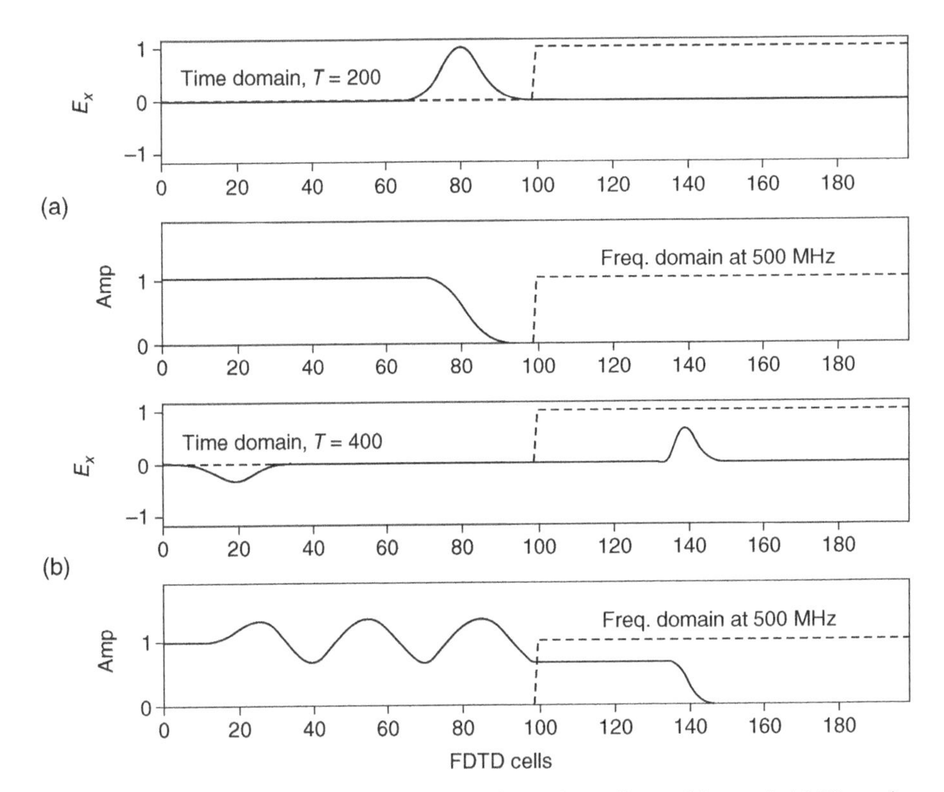

Figure 2.1 Simulation of a pulse striking a dielectric medium with $\varepsilon_r = 4$. (a) The pulse after 200 time steps. Notice that the Fourier amplitude is 1 in that part of the space where the pulse has traveled, but 0 elsewhere. (b) After 400 time steps, the pulse has struck the medium, part of it has been transmitted and the other part reflected. The Fourier amplitude in the medium is 0.667, which is the percentage that has been transmitted.

2.3 FREQUENCY-DEPENDENT MEDIA

The dielectric constant and conductivity of most media vary at different frequencies. The pulses we have been using as sources in Chapters 1 and 2 contain a spectrum of frequencies. In order to simulate frequency-dependent material, we will need a way to account for this. One of the most significant developments in the FDTD method is the technique to simulate frequency-dependent materials (3).

We will start with a very simple example to illustrate the ideas. Suppose we have a medium whose dielectric constant and conductivity vary over the

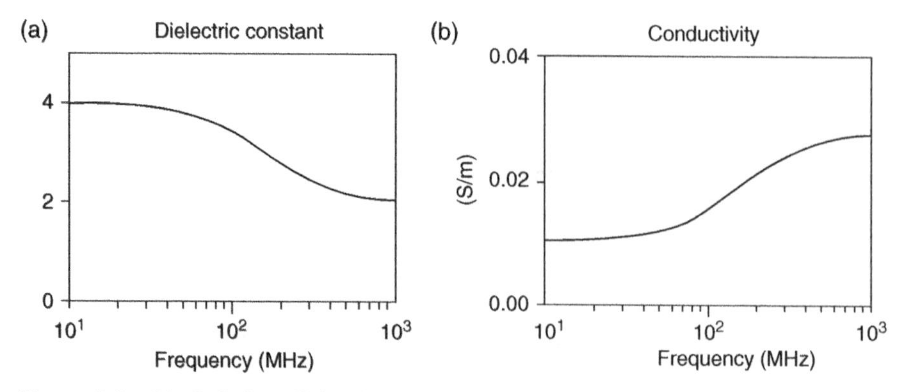

Figure 2.2 (a) Relative dielectric constant and (b) conductivity as functions of frequency for a Debye medium with the following properties: $\varepsilon_r = 2$, $\sigma = 0.01$, $\chi_1 = 2$, and $\tau = 0.001$ μs.

frequency range of 10–1000 MHz as shown (Fig. 2.2). A material like this can be adequately represented by the following formulation:

$$\varepsilon_r^*(\omega) = \varepsilon_r + \frac{\sigma}{j\omega \cdot \varepsilon_0} + \frac{\chi_1}{1 + j\omega \cdot \tau}. \tag{2.16}$$

This is referred to as the Debye formulation. In this formulation, there is a dielectric constant ε_r and a conductivity σ, but there is also a frequency-dependent term. The following parameters represent the medium of Fig. 2.2:

$$\varepsilon_r = 2, \quad \sigma = 0.01, \quad \chi_1 = 2, \quad \tau = 0.001 \mu\text{s}$$

In order to simulate this medium in FDTD, Eq. (2.16) must be put into the sampled time domain. Let us define the last term times the $\boldsymbol{E}$ field as

$$\boldsymbol{S}(\omega) = \frac{\chi_1}{1 + j\omega \cdot \tau} \boldsymbol{E}(\omega). \tag{2.17}$$

The inverse Fourier transform of the Debye term is $(\chi_1/\tau) \cdot e^{-(t/\tau)} \cdot u(t)$, where $u(t)$ is the rectangular function, which is 0 for $t < 0$ and 1 thereafter. (Recall that implicit to FDTD is the fact that all functions are causal, that is, 0 for $t < 0$, because the computer programs initialize the field values to 0.) Equation (2.17) in the frequency domain becomes the convolution

$$\boldsymbol{S}(t) = \frac{\chi_1}{\tau} \int_0^t e^{-(t-t')/\tau} \cdot \boldsymbol{E}(t') \, dt'$$

in the time domain. We now have to approximate this as a summation in the sampled time domain:

$$
\begin{aligned}
\boldsymbol{S}^n &= \chi_1 \cdot \frac{\Delta t}{\tau} \sum_{i=0}^{n} \mathrm{e}^{-(\Delta t)\cdot(n-i)/\tau} \cdot \boldsymbol{E}^i \\
&= \chi_1 \cdot \frac{\Delta t}{\tau} \left[\boldsymbol{E}^n + \sum_{i=0}^{n-1} \mathrm{e}^{-(\Delta t)\cdot(n-i)/\tau} \cdot \boldsymbol{E}^i \right].
\end{aligned}
\tag{2.18}
$$

Notice that

$$
\begin{aligned}
\boldsymbol{S}^{n-1} &= \chi_1 \cdot \frac{\Delta t}{\tau} \sum_{i=0}^{n-1} \mathrm{e}^{-(\Delta t)\cdot(n-1-i)/\tau} \cdot \boldsymbol{E}^i \\
&= \chi_1 \cdot \mathrm{e}^{\Delta t/\tau} \frac{\Delta t}{\tau} \sum_{i=0}^{n-1} \mathrm{e}^{-(\Delta t)\cdot(n-i)/\tau} \cdot \boldsymbol{E}^i.
\end{aligned}
$$

Substituting this value into Eq. (2.18) above gives

$$
\boldsymbol{S}^n = \mathrm{e}^{-\Delta t/\tau} \cdot \boldsymbol{S}^{n-1} + \chi_1 \cdot \frac{\Delta t}{\tau} \cdot \boldsymbol{E}^n. \tag{2.19}
$$

Similar to the way we handled the lossy dielectric, we can write

$$
\begin{aligned}
\boldsymbol{D}^n &= \varepsilon_r \boldsymbol{E}^n + \boldsymbol{I}^n + \boldsymbol{S}^n \\
&= \varepsilon_r \boldsymbol{E}^n + \left(\frac{\sigma \cdot \Delta t}{\varepsilon_0} \boldsymbol{E}^n + \boldsymbol{I}^{n-1} \right) + \left(\chi_1 \cdot \frac{\Delta t}{\tau} \boldsymbol{E}^n + \mathrm{e}^{-\Delta t/\tau} \cdot \boldsymbol{S}^{n-1} \right),
\end{aligned}
\tag{2.20}
$$

and solving for $\boldsymbol{E}^n$:

$$
\boldsymbol{E}^n = \frac{\boldsymbol{D}^n - \boldsymbol{I}^{n-1} - \mathrm{e}^{-\Delta t/\tau} \cdot \boldsymbol{S}^{n-1}}{\varepsilon_r + \dfrac{\sigma \cdot \Delta t}{\varepsilon_0} + \chi_1 \cdot \dfrac{\Delta t}{\tau}}, \tag{2.21a}
$$

$$
\boldsymbol{I}^n = \boldsymbol{I}^{n-1} + \frac{\sigma \cdot \Delta t}{\varepsilon_0} \boldsymbol{E}^n, \tag{2.21b}
$$

$$
\boldsymbol{S}^n = \mathrm{e}^{-\Delta t/\tau} \cdot \boldsymbol{S}^{n-1} + \chi_1 \cdot \frac{\Delta t}{\tau} \boldsymbol{E}^n. \tag{2.21c}
$$

This formulation is implemented in one dimension, again in the same orientation used in Chapter 1, by the following Python code:

```
dx[k] = dx[k] + 0.5 * (hy[k - 1] - hy[k])                    (2.22a)
ex[k] = gax[k] * (dx[k] - ix[k] - del_exp * sx[k])           (2.22b)
ix[k] = ix[k] + gbx[k] * ex[k]                               (2.22c)
sx[k] = del_exp * sx[k] + gcx[k] * ex[k]                     (2.22d)
hy[k] = hy[k] + 0.5 * (ex[k] - ex[k + 1])                    (2.22e)
```

where

```
gax[k] = 1/(epsr + (sigma * dt/epsz) + chi * dt/tau)   (2.23a)
gbx[k] = sigma * dt/epsz                               (2.23b)
gcx[k] = chi * dt/tau                                  (2.23c)
```

and

```
del_exp = exp( - dt/tau).
```

Once again, note that everything concerning the medium is contained in Eq. (2.22b), (2.22c), and (2.22d); Eq. (2.22a) and (2.22e), the calculation of the flux density and the magnetic field, respectively, are unchanged.

The program fd1d_2_3.py calculates the frequency-domain amplitude and phase for three frequencies. Figure 2.3 shows a simulation of a pulse going into a frequency-dependent dielectric material with the properties

$$\varepsilon_r = 2, \quad \sigma = 0.01, \quad \chi_1 = 2, \quad \tau = 0.001\,\mu\text{s}.$$

This set of parameters has the following effective dielectric constants and conductivities at the three frequencies:

Frequency (MHz)	ε_r	σ(S/m)
50	3.82	0.012
200	2.78	0.021
500	2.18	0.026

Notice that the Fourier amplitude attenuates more rapidly at 200 MHz than at 50 MHz, and more rapidly still at 500 MHz. This is because the conductivity is higher at these frequencies. At the same time, the higher relative dielectric constant at the lower frequencies means that the amplitude just inside the medium is smaller.

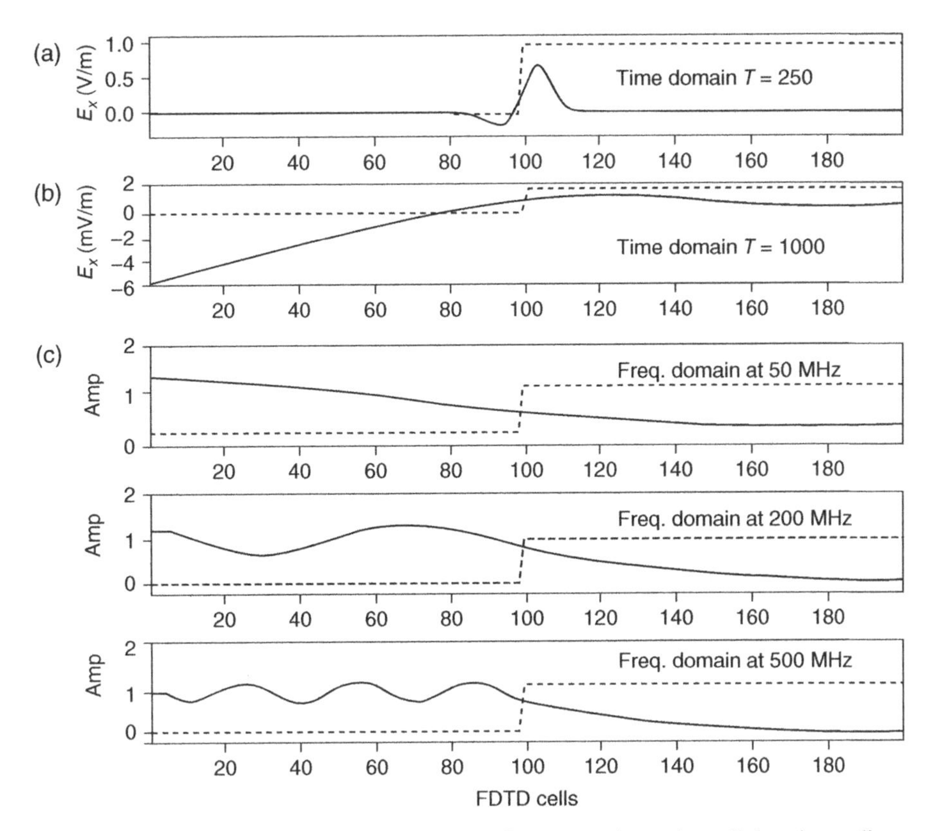

Figure 2.3 Simulation of a pulse striking a frequency-dependent dielectric medium (Debye medium) with the following properties: $\varepsilon_r = 2$, $\sigma = 0.01$, $\chi_1 = 2$, and $\tau = 0.001$ μs. (a) After 250 time steps, the pulse has struck the medium, and part of it has been transmitted and part reflected. (b) After 1000 steps the pulse has penetrated into the medium, but has spread. (c) Notice the different percentages of transmittance and the different rates of attenuation within the medium at each frequency due to the different effective dielectric constants and conductivities.

2.3.1 Auxiliary Differential Equation Method

We could have taken a somewhat different approach to the simulation of the dispersive medium described by Eq. (2.16): Namely, the auxiliary differential equation (ADE) method (4, 5). Let us look at Eq. (2.17), but rewrite it as

$$(1 + j\omega \cdot \tau)\boldsymbol{S}(\omega) = \chi_1 \cdot \boldsymbol{E}(\omega). \tag{2.24}$$

Once again, we must find a way to take this to the discrete time domain for implementation in the FDTD formulation. We will start by going to the continuous time domain, and Eq. (2.24) becomes

$$s(t) + \tau \frac{\mathrm{d}s(t)}{\mathrm{d}t} = \chi_1 \cdot e(t). \tag{2.25}$$

In the sampled time domain, this becomes

$$\frac{\boldsymbol{S}^n + \boldsymbol{S}^{n-1}}{2} + \tau \frac{\boldsymbol{S}^n - \boldsymbol{S}^{n-1}}{\Delta t} = \chi_1 \cdot \boldsymbol{E}^n.$$

Notice that we approximate the $s(t)$ term over two time steps. We did this because we needed two time steps to approximate the derivatives. As before, we solve for $\boldsymbol{S}^n$:

$$\boldsymbol{S}^n = \frac{\left(1 - \frac{\Delta t}{2 \cdot \tau}\right)}{\left(1 + \frac{\Delta t}{2 \cdot \tau}\right)} \boldsymbol{S}^{n-1} + \frac{\frac{\Delta t}{\tau} \cdot \chi_1}{\left(1 + \frac{\Delta t}{2 \cdot \tau}\right)} \boldsymbol{E}^n. \tag{2.26}$$

We can use this instead of Eq. (2.19) to calculate $\boldsymbol{E}^n$ in Eq. (2.21). How can we be sure these will give equivalent answers? You are probably familiar with the following approximations:

$$1 - \delta \cong \mathrm{e}^{-\delta} \quad \text{if } \delta \ll 1,$$

$$\frac{1}{1 + \delta} \cong \mathrm{e}^{-\delta} \quad \text{if } \delta \ll 1.$$

Putting the two together gives

$$\frac{1 - \delta}{1 + \delta} \cong \mathrm{e}^{-2\delta} \quad \text{if } \delta \ll 1.$$

In this case

$$\delta = \frac{\Delta t}{2 \cdot \tau},$$

so we have

$$\frac{\left(1 - \frac{\Delta t}{2 \cdot \tau}\right)}{\left(1 + \frac{\Delta t}{2 \cdot \tau}\right)} \cong \mathrm{e}^{-\Delta t/\tau}. \tag{2.27}$$

We now have only the following question: How do we know that $\Delta t/\tau$ is small enough? Recall that the cell size has to be small enough to get about 10 points per wavelength for the smallest wavelength in the simulation. This is a similar situation. If the medium that we are trying to simulate has a Debye term with a time constant of τ, we must be sure that our time steps are small compared to τ, say $\Delta t \leq \tau/10$. This ensures that Eq. (2.27) is a fairly good approximation.

PROBLEM SET 2.3

1. The program fd1d_2_3.py implements the frequency-dependent formulation. Get this program running and repeat the results of Fig. 2.3.

2.4 FORMULATION USING Z TRANSFORMS

If you have been studying the Z transform theory in Appendix A, or if you are already familiar with Z transforms, you will now know the advantage of using Z transforms for the FDTD formulation of frequency-dependent media (6). Returning to the problem of calculating $\boldsymbol{E}$ in a Debye medium, we begin with our frequency domain equations:

$$\boldsymbol{D}(\omega) = \left(\varepsilon_r + \frac{\sigma}{j\omega \cdot \varepsilon_0} + \frac{\chi_1}{1 + j\omega \cdot \tau}\right)\boldsymbol{E}(\omega).$$

We can avoid dealing with troublesome convolution integrals in the time domain because we will go immediately to the Z domain:

$$\boldsymbol{D}(z) = \varepsilon_r \boldsymbol{E}(z) + \frac{\sigma \cdot \dfrac{\Delta t}{\varepsilon_0}}{1 - z^{-1}}\boldsymbol{E}(z) + \frac{\chi_1 \cdot \dfrac{\Delta t}{\tau}}{1 - \mathrm{e}^{-\Delta t/\tau} \cdot z^{-1}}\boldsymbol{E}(z). \tag{2.28}$$

Notice that the factor Δt, which is the time step, had to be added and shows up in the last two terms in going from the time domain to the Z domain. Similar to what we did earlier, we will define some auxiliary parameters:

$$\boldsymbol{I}(z) = \frac{\sigma \cdot \dfrac{\Delta t}{\varepsilon_0}}{1 - z^{-1}}\boldsymbol{E}(z) = z^{-1} \cdot \boldsymbol{I}(z) + \sigma \cdot \frac{\Delta t}{\varepsilon_0}\boldsymbol{E}(z), \tag{2.29a}$$

$$\boldsymbol{S}(z) = \frac{\chi_1 \cdot \dfrac{\Delta t}{\tau}}{1 - \mathrm{e}^{-\Delta t/\tau} \cdot z^{-1}} \cdot \boldsymbol{E}(z) = \mathrm{e}^{-\Delta t/\tau} \cdot z^{-1} \cdot \boldsymbol{S}(z) + \chi_1 \cdot \frac{\Delta t}{\tau} \cdot \boldsymbol{E}(z). \tag{2.29b}$$

Equation (2.28) then becomes

$$\begin{aligned} \boldsymbol{D}(z) = {} & \varepsilon_r \boldsymbol{E}(z) + z^{-1} \cdot \boldsymbol{I}(z) + \sigma \cdot \frac{\Delta t}{\varepsilon_0} \cdot \boldsymbol{E}(z) \\ & + \mathrm{e}^{-\Delta t/\tau} \cdot z^{-1} \cdot \boldsymbol{S}(z) + \chi_1 \cdot \frac{\Delta t}{\tau} \cdot \boldsymbol{E}(z), \end{aligned} \tag{2.30}$$

from which we can solve for $\boldsymbol{E}(z)$ by

$$\boldsymbol{E}(z)=\frac{\boldsymbol{D}(z)-z^{-1}\cdot \boldsymbol{I}(z)-\mathrm{e}^{-\Delta t/\tau}\cdot z^{-1}\cdot \boldsymbol{S}(z)}{\varepsilon_r+\sigma\cdot \dfrac{\Delta t}{\varepsilon_0}+\chi_1\cdot \dfrac{\Delta t}{\tau}}. \tag{2.31}$$

The advantage of the Z transform formulation is that to get to the sampled time domain, replace $\boldsymbol{E}(z)$ with $\boldsymbol{E}^n$, $z^{-1}\cdot \boldsymbol{E}(z)$ with $\boldsymbol{E}^{n-1}$, and similarly replace the other parameters in Eq. (2.31), (2.29a), and (2.29b). What you get is

$$\boldsymbol{E}^n=\frac{\boldsymbol{D}^n-\boldsymbol{I}^{n-1}-\mathrm{e}^{-\Delta t/\tau}\cdot \boldsymbol{S}^{n-1}}{\varepsilon_r+\dfrac{\sigma\cdot \Delta t}{\varepsilon_0}+\chi_1\cdot \dfrac{\Delta t}{\tau}}, \tag{2.32a}$$

$$\boldsymbol{I}^n=\boldsymbol{I}^{n-1}+\sigma\cdot \frac{\Delta t}{\varepsilon_0}\cdot \boldsymbol{E}^n, \tag{2.32b}$$

$$\boldsymbol{S}^n=\mathrm{e}^{-\Delta t/\tau}\cdot \boldsymbol{S}^{n-1}+\chi_1\cdot \frac{\Delta t}{\tau}\cdot \boldsymbol{E}^n, \tag{2.32c}$$

which is *exactly* what we got in the previous section. The difference is that we avoided doing anything with integrals and their approximations. As we move to formulations that are more complicated, the advantage of the Z transform will become evident.

2.4.1 Simulation of Unmagnetized Plasma

In this section, we will demonstrate the versatility of the methods we have learned in this chapter by simulating a medium completely different from the media we have been working with so far. The permittivity of unmagnetized plasmas is given as (7)

$$\varepsilon^*(\omega)=1+\frac{\omega_p^2}{\omega(j\nu_c-\omega)}, \tag{2.33}$$

where

$$\omega_p=2\pi\cdot f_p;$$

f_p= the plasma frequency;
ν_c= the electron collision frequency.

Using partial fraction expansion, Eq. (2.33) can be written as

$$\varepsilon^*(\omega)=1+\frac{\dfrac{\omega_p^2}{\nu_c}}{j\omega}-\frac{\dfrac{\omega_p^2}{\nu_c}}{\nu_c+j\omega}. \tag{2.34}$$

This is the value we will use for the complex dielectric constant in Eq. (2.3b). Notice that the form resembles Eq. (2.16), the expression for a lossy material with a Debye term.

There are several ways of approaching this problem, but let us start by taking the Z transforms of Eq. (2.34) to obtain

$$\varepsilon^*(z) = \frac{1}{\Delta t} + \frac{\frac{\omega_p^2}{\nu_c}}{1 - z^{-1}} - \frac{\frac{\omega_p^2}{\nu_c}}{1 - e^{-\nu_c \cdot \Delta t} z^{-1}}. \tag{2.35}$$

By the convolution theorem, the Z transform of Eq. (2.3b) is

$$\boldsymbol{D}(z) = \varepsilon^*(z) \cdot \boldsymbol{E}(z) \cdot \Delta t. \tag{2.36}$$

By inserting Eq. (2.35) into Eq. (2.36), we obtain

$$\begin{aligned}\boldsymbol{D}(z) &= \boldsymbol{E}(z) + \frac{\omega_p^2 \cdot \Delta t}{\nu_c} \left[\frac{1}{1 - z^{-1}} - \frac{1}{1 - e^{-\nu_c \cdot \Delta t} z^{-1}} \right] \boldsymbol{E}(z) \\ &= \boldsymbol{E}(z) + \frac{\omega_p^2 \cdot \Delta t}{\nu_c} \left[\frac{(1 - e^{-\nu_c \cdot \Delta t}) z^{-1}}{1 - (1 + e^{-\nu_c \cdot \Delta t}) z^{-1} + e^{-\nu_c \cdot \Delta t} z^{-2}} \right] \boldsymbol{E}(z).\end{aligned}$$

Notice that we cross multiplied the term in the brackets. An auxiliary term will be defined as

$$\boldsymbol{S}(z) = \frac{\omega_p^2 \cdot \Delta t}{\nu_c} \left[\frac{(1 - e^{-\nu_c \cdot \Delta t})}{1 - (1 + e^{-\nu_c \cdot \Delta t}) z^{-1} + e^{-\nu_c \cdot \Delta t} z^{-2}} \right] \boldsymbol{E}(z).$$

$\boldsymbol{E}(z)$ can be solved for by

$$\boldsymbol{E}(z) = \boldsymbol{D}(z) - z^{-1} \cdot \boldsymbol{S}(z), \tag{2.37a}$$

$$\boldsymbol{S}(z) = \left(1 + e^{-\nu_c \cdot \Delta t}\right) z^{-1} \cdot \boldsymbol{S}(z) - e^{-\nu_c \cdot \Delta t} z^{-2} \cdot \boldsymbol{S}(z) + \frac{\omega_p^2 \cdot \Delta t}{\nu_c} \left(1 - e^{-\nu_c \cdot \Delta t}\right) \cdot \boldsymbol{E}(z). \tag{2.37b}$$

Therefore, the FDTD simulation, in one dimension with the usual orientation, becomes

```
ex[k]  =  dx[k]  -  sx[k]                              (2.38a)
sx[k] = (1 + exp(-vc[k] * dt)) * sxm1[k] - \
        exp( - vc[k] * dt) *sxm2[k] + \                (2.38b)
        (omega * * 2 * dt/vc[k]) * \
        (1-exp(-vc[k] * dt)) * ex[k]
```

$$\texttt{sxm2[k]} = \texttt{sxm1[k]} \tag{2.38c}$$

$$\texttt{sxm1[k]} = \texttt{sx[k]} \tag{2.38d}$$

Notice from Eq. (2.38b) that we need the two previous values of S in our calculation. This is accomplished by Eq. (2.38c) and (2.38d).

We will do a simulation of a pulse propagating in free space that comes upon plasma, which is a very interesting medium. At relatively low frequencies (below the plasma resonance frequency f_p), it looks like a metal, and at higher frequencies (above the plasma resonance frequency f_p), it becomes transparent. The following simulation uses the properties of silver: $\nu_p = 57$ THz, $f_p = 2000$ THz. Of course, since we are simulating much higher frequencies, we will need a much smaller cell size. In the course of this problem, it will be necessary to simulate EM waves of 4000 THz. At this frequency, the free space wavelength is

$$\lambda = \frac{3 \times 10^8}{4 \times 10^{15}} = 0.75 \times 10^{-7}\text{m}.$$

In following our rule of thumb of at least 10 points per wavelength, a cell size of 5 nm, that is, $\Delta x = 5 \times 10^{-9}$m, will be used. Figure 2.4 shows our first simulation at 500 THz, well below the plasma frequency. For this frequency, $\Delta x = 1 \times 10^{-8}$ m was chosen. The incident pulse is a sine wave inside a Gaussian envelope. Notice that the pulse interacts with the plasma almost as if the plasma were a metal barrier—the pulse is almost completely reflected. Figure 2.5 is a

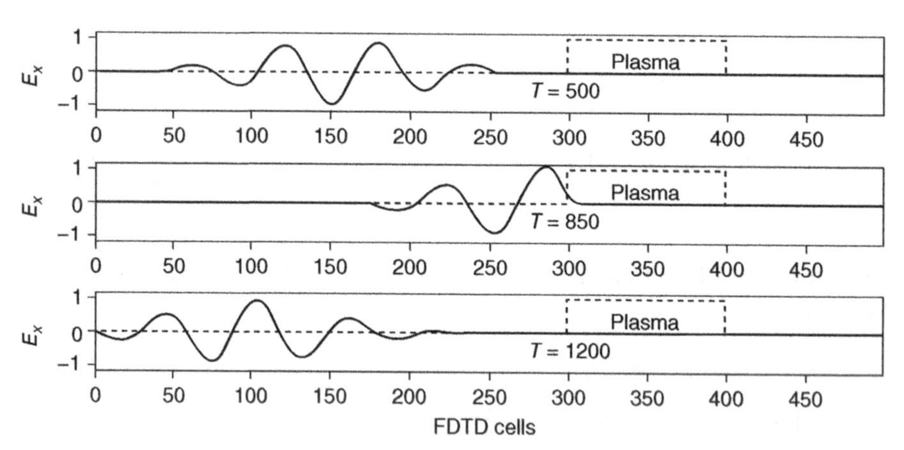

Figure 2.4 Simulation of a wave propagating in free space and striking a plasma medium. The plasma has the properties of silver: $f_p = 2000$ THz and $\nu_c = 57$ THz. The propagating wave has a center frequency of 500 THz. After 1200 time steps, the wave has been completely reflected by the plasma.

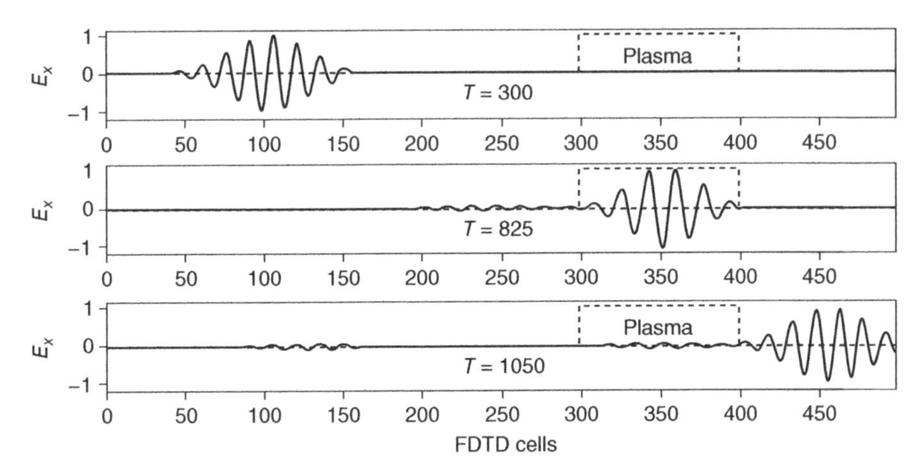

Figure 2.5 Simulation of a wave propagating in free space and striking a plasma medium. The plasma has the properties of silver: f_p = 2000 THz and ν_c = 57 THz. The propagating wave has a center frequency of 4000 THz. After 1050 time steps, the majority of the signal has passed through the plasma.

similar simulation at 4000 THz, well above the plasma frequency. A small portion of it is reflected, but the majority of the pulse passes right through.

PROBLEM SET 2.4

1. Modify the program fd1d_2_3.py to simulate plasma and duplicate the results of Fig. 2.4 and Fig. 2.5. This is much easier than it might look. First, change your cell size to 10 nm and 5 nm, respectively. Create a ν_c array with a very small number chosen for free space, and ν_c = 57 THz for the plasma. Then, replace the calculation of the ***E*** field for a lossy Debye medium with that of Eq. (2.38). After doing the simulations at 500 and 4000 THz, repeat at 2000 THz. What happens?
2. Repeat Problem 2.4.1, but use a narrow Gaussian pulse as your input and calculate the frequency response at 500, 2000, and 4000 THz. Does the result look the way you would expect? Particularly at 2000 THz?

2.5 FORMULATING A LORENTZ MEDIUM

The Debye model describes a single-pole frequency dependence. We now move to the next level, which is a two-pole dependence referred to as the *Lorentz formulation*:

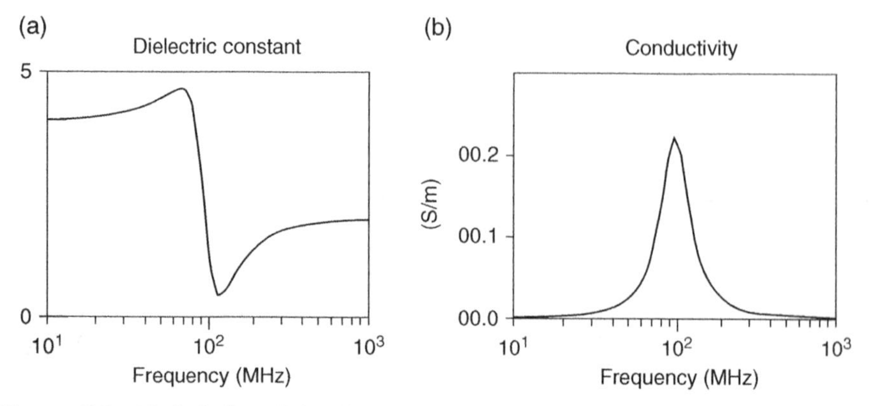

Figure 2.6 (a) Relative dielectric constant and (b) conductivity as functions of frequency for a Lorentz medium with the following properties: $\varepsilon_r = 2$, $\varepsilon_1 = 2$, $f_0 = 100$ MHz, and $\delta_0 = 0.25$.

$$\varepsilon_r^*(\omega) = \varepsilon_r + \frac{\varepsilon_1}{1 + j \cdot 2\delta_0 \left(\frac{\omega}{\omega_0}\right) - \left(\frac{\omega}{\omega_0}\right)^2}. \tag{2.39}$$

Figure 2.6 is a graph of the dielectric constant and conductivity of a material with the following Lorentz parameters: $\varepsilon_r = 2$, $\varepsilon_1 = 2$, $f_0 = 100$ MHz, $(\omega_0 = 2\pi f_0)$, and $\delta_0 = 0.25$. To simulate a Lorentz medium in the FDTD formulation, we first put Eq. (2.39) into Eq. (2.3b),

$$\begin{aligned} \boldsymbol{D}(\omega) &= \varepsilon_r \boldsymbol{E}(\omega) + \frac{\varepsilon_1}{1 + j \cdot 2\delta_0 \left(\frac{\omega}{\omega_0}\right) - \left(\frac{\omega}{\omega_0}\right)^2} \boldsymbol{E}(\omega) \\ &= \varepsilon_r \boldsymbol{E}(\omega) + \boldsymbol{S}(\omega), \end{aligned} \tag{2.40}$$

where we have defined again an auxiliary term,

$$\boldsymbol{S}(\omega) = \frac{\omega_0^2 \varepsilon_1}{\omega_0^2 + j \cdot 2\delta_0 \omega_0 \omega + (j\omega)^2} \boldsymbol{E}(\omega).$$

We will use the ADE method to move this to the time domain. Start by rewriting it in the following manner:

$$\left[\omega_0^2 + j \cdot 2\delta_0 \omega_0 \omega + (j\omega)^2\right] \boldsymbol{S}(\omega) = \omega_0^2 \varepsilon_1 \boldsymbol{E}(\omega).$$

Finally, we proceed to the finite difference approximations:

$$\omega_0^2 \boldsymbol{S}^{n-1} + 2\delta_0\omega_0 \frac{\boldsymbol{S}^n - \boldsymbol{S}^{n-2}}{2 \cdot \Delta t} + \frac{\boldsymbol{S}^n - 2\boldsymbol{S}^{n-1} + \boldsymbol{S}^{n-2}}{(\Delta t)^2} = \omega_0^2 \varepsilon_1 \boldsymbol{E}^{n-1}.$$

A few things are worth noting: the second-order derivative generated a second-order differencing:

$$\frac{\mathrm{d}^2 s(t)}{\mathrm{d}t^2} \cong \frac{\boldsymbol{S}^n - 2\boldsymbol{S}^{n-1} + \boldsymbol{S}^{n-2}}{(\Delta t)^2}.$$

The first-order derivative is taken over two time steps instead of one:

$$\frac{\mathrm{d}s(t)}{\mathrm{d}t} \cong \frac{\boldsymbol{S}^n - \boldsymbol{S}^{n-2}}{2 \cdot \Delta t}.$$

This was done because the second-order derivative spanned two time steps. Next, we solve for the newest value of $\boldsymbol{S}^n$:

$$\boldsymbol{S}^n \left[\frac{\delta_0\omega_0}{\Delta t} + \frac{1}{(\Delta t)^2}\right] + \boldsymbol{S}^{n-1}\left[\omega_0^2 - \frac{2}{(\Delta t)^2}\right] + \boldsymbol{S}^{n-2}\left[-\frac{\delta_0\omega_0}{\Delta t} + \frac{1}{(\Delta t)^2}\right] = \omega_0^2 \varepsilon_1 \boldsymbol{E}^{n-1},$$

$$\boldsymbol{S}^n = -\frac{\left[\omega_0^2 - \frac{2}{(\Delta t)^2}\right]}{\left[\frac{\delta_0\omega_0}{\Delta t} + \frac{1}{(\Delta t)^2}\right]}\boldsymbol{S}^{n-1} - \frac{\left[\frac{1}{(\Delta t)^2} - \frac{\delta_0\omega_0}{\Delta t}\right]}{\left[\frac{\delta_0\omega_0}{\Delta t} + \frac{1}{(\Delta t)^2}\right]}\boldsymbol{S}^{n-2} + \frac{\omega_0^2\varepsilon_1}{\left[\frac{\delta_0\omega_0}{\Delta t} + \frac{1}{(\Delta t)^2}\right]}\boldsymbol{E}^{n-1},$$

$$\boldsymbol{S}^n = \frac{\left[2 - (\Delta t)^2\omega_0^2\right]}{(1 + \Delta t \cdot \delta_0\omega_0)}\boldsymbol{S}^{n-1} - \frac{(1 - \Delta t \cdot \delta_0\omega_0)}{(1 + \Delta t \cdot \delta_0\omega_0)}\boldsymbol{S}^{n-2} + \frac{(\Delta t)^2\omega_0^2\varepsilon_1}{(1 + \Delta t \cdot \delta_0\omega_0)}\boldsymbol{E}^{n-1}. \tag{2.41}$$

Now, we will return to Eq. (2.40) and take it to the sampled time domain

$$\boldsymbol{D}^n = \varepsilon_r \boldsymbol{E}^n + \boldsymbol{S}^n,$$

which we rewrite as

$$\boldsymbol{E}^n = \frac{\boldsymbol{D}^n - \boldsymbol{S}^n}{\varepsilon_r}.$$

Notice that we already have a solution for $\boldsymbol{S}^n$ in Eq. (2.41). Also, because we need only the previous value of $\boldsymbol{E}$, that is, $\boldsymbol{E}^{n-1}$, we are done.

Let us try a different approach. Go back to Eq. (2.39). An alternative form of the Lorentz formulation is

$$S(\omega) = \frac{\gamma\beta}{(\alpha^2 + \beta^2) + j\omega 2\alpha - \omega^2} \varepsilon_1 E(\omega), \tag{2.42}$$

where

$$\gamma = \frac{\omega_0}{\sqrt{1 - \delta_0^2}},$$

$$\alpha = \delta_0 \omega_0,$$

$$\beta = \omega_0 \sqrt{1 - \delta_0^2}.$$

At this point, we can go to Table A.1 and look up the corresponding Z transform, which is

$$S(z) = \frac{e^{-\alpha \cdot \Delta t} \cdot \sin(\beta \cdot \Delta t) \cdot \Delta t \cdot z^{-1}}{1 - 2e^{-\alpha \cdot \Delta t} \cdot \cos(\beta \cdot \Delta t) \cdot z^{-1} + e^{-2\alpha \cdot \Delta t} \cdot z^{-2}} \gamma \varepsilon_1 E(z).$$

The corresponding sampled time-domain equation is

$$S^n = 2e^{-\alpha \cdot \Delta t} \cdot \cos(\beta \cdot \Delta t) \cdot S^{n-1} - e^{-2\alpha \cdot \Delta t} \cdot S^{n-2} + e^{-\alpha \cdot \Delta t} \sin(\beta \cdot \Delta t) \cdot \Delta t \cdot \gamma \varepsilon_1 \cdot E^{n-1}. \tag{2.43}$$

If both methods are correct, they should yield the same answers, which means the corresponding terms in Eq. (2.41) and Eq. (2.43) should be the same. Let us see if that is true.

S^{n-2} term: We saw earlier that

$$e^{-2\delta_0 \omega_0 \cdot \Delta t} \cong \frac{1 - \delta_0 \omega_0 \cdot \Delta t}{1 + \delta_0 \omega_0 \cdot \Delta t}.$$

E^{n-1} term: If $\beta \cdot \Delta t \ll 1$, then $\sin(\beta \cdot \Delta t) \cong \beta \cdot \Delta t$, and

$$\begin{aligned} e^{-\alpha \cdot \Delta t} \sin(\beta \cdot \Delta t) \cdot \gamma \cdot \varepsilon_1 \cdot \Delta t &= \frac{\beta \cdot \Delta t}{1 + \alpha \cdot \Delta t} \gamma \cdot \varepsilon_1 \cdot \Delta t \\ &= \frac{\omega_0 \sqrt{1 - \delta_0^2} \cdot \Delta t}{1 + \delta_0 \omega_0 \cdot \Delta t} \cdot \frac{\omega_0}{\sqrt{1 - \delta_0^2}} \cdot \varepsilon_1 \cdot \Delta t = \frac{\varepsilon_1 \cdot \omega_0^2 \cdot (\Delta t)^2}{1 + \delta_0 \omega_0 \cdot \Delta t}. \end{aligned}$$

S^{n-1} term: By using a Taylor series expansion of the cosine function, we get

$$2e^{-\alpha\cdot\Delta t}\cos(\beta\cdot\Delta t) = \frac{2}{1+\alpha\cdot\Delta t}\left[1-\frac{(\beta\cdot\Delta t)^2}{2}\right]$$
$$= \frac{2-(\beta\cdot\Delta t)^2}{1+\alpha\cdot\Delta t} = \frac{2-\omega_0^2\cdot\left(1-\delta_0^2\right)(\Delta t)^2}{1+\delta_0\omega_0\cdot\Delta t}.$$

The last step is perhaps our weakest one. It depends on δ_0 being small enough compared to one that δ_0^2 will be negligible.

2.5.1 Simulation of Human Muscle Tissue

We will conclude with the simulation of human muscle tissue. Muscle tissue can be adequately simulated over a frequency range of about two decades with the following formulation (8):

$$\varepsilon_r^*(\omega) = \varepsilon_r + \frac{\sigma}{j\omega\cdot\varepsilon_0} + \varepsilon_1\frac{\omega_0}{\left(\omega_0^2+\alpha^2\right)+j\omega\cdot 2\alpha-\omega^2}. \tag{2.44}$$

(Problem 2.5.2 addresses how one determines the specific parameters, but we will not be concerned with that here.) Inserting Eq. (2.44) into Eq. (2.3b) and taking the Z transform, we get

$$\begin{aligned} D(z) &= \varepsilon_r E(z) + \frac{\sigma\cdot\frac{\Delta t}{\varepsilon_0}}{1-z^{-1}}E(z) \\ &+ \varepsilon_1\frac{e^{-\alpha\cdot\Delta t}\sin(\omega_0\cdot\Delta t)\cdot\Delta t\cdot z^{-1}}{1-2e^{-\alpha\cdot\Delta t}\cos(\omega_0\cdot\Delta t)\cdot z^{-1}+e^{-2\alpha\cdot\Delta t}z^{-2}}E(z). \end{aligned} \tag{2.45}$$

We will define two auxiliary parameters:

$$I(z) = \frac{\sigma\cdot\frac{\Delta t}{\varepsilon_0}}{1-z^{-1}}E(z), \tag{2.46a}$$

$$S(z) = \varepsilon_1\frac{e^{-\alpha\cdot\Delta t}\sin(\omega_0\cdot\Delta t)\cdot\Delta t}{1-2e^{-\alpha\cdot\Delta t}\cos(\omega_0\cdot\Delta t)\cdot z^{-1}+e^{-2\alpha\cdot\Delta t}z^{-2}}E(z). \tag{2.46b}$$

Now, Eq. (2.45) becomes

$$D(z) = \varepsilon_r E(z) + I(z) + z^{-1}\cdot S(z). \tag{2.47}$$

Once we have calculated $\boldsymbol{E}(z)$, $\boldsymbol{I}(z)$ and $\boldsymbol{S}(z)$ can be calculated from Eq. (2.46a) and (2.46b). Rearranging these equations and Eq. (2.47) leads to:

$$\boldsymbol{E}(z) = \frac{\boldsymbol{D}(z) - z^{-1} \cdot \boldsymbol{I}(z) - z^{-1} \cdot \boldsymbol{S}(z)}{\varepsilon_r + \sigma \cdot \dfrac{\Delta t}{\varepsilon_0}}, \tag{2.48a}$$

$$\boldsymbol{I}(z) = z^{-1} \cdot \boldsymbol{I}(z) + \frac{\sigma \cdot \Delta t}{\varepsilon_0} \boldsymbol{E}(z), \tag{2.48b}$$

$$\begin{aligned} \boldsymbol{S}(z) = {} & 2e^{-\alpha \cdot \Delta t} \cos(\omega_0 \cdot \Delta t) \cdot z^{-1} \cdot \boldsymbol{S}(z) \\ & - e^{-2\alpha \cdot \Delta t} \cdot z^{-2} \cdot \boldsymbol{S}(z) + \varepsilon_1 \cdot e^{-\alpha \cdot \Delta t} \sin(\omega_0 \cdot \Delta t) \cdot (\Delta t) \cdot \boldsymbol{E}(z). \end{aligned} \tag{2.48c}$$

PROBLEM SET 2.5

1. Modify fd1d_2_3.py to simulate a Lorentz medium with the properties used in Fig. 2.6. Calculate the Fourier amplitudes at 50, 100, and 200 MHz. Considering Fig. 2.6, do you get the results you expect?
2. Quantify the results of Problem 2.5.1 by comparing them with analytic calculations of the reflection coefficient and transmission coefficient at the three frequencies using Appendix 1.A.
3. FDTD simulation has been used extensively to model the effects of electromagnetic radiation on human tissue for both safety (9) and for therapeutic applications (10). Human muscle tissue is highly frequency dependent. Table 2.1 shows how the dielectric constant and the conductivity of muscle vary with frequency (11). Other tissues display similar frequency dependence (8).

   Write a program to calculate the values of ε_r, σ, χ_1, and t_0 that would give an adequate Debye representation of the data in Table 2.1. You probably will not be able to fit the values exactly. (Hint: Do *not* try for a purely analytic solution; do it by trial and error.)

TABLE 2.1 Properties of Human Muscle

Frequency (MHz)	Dielectric Constant	Conductivity (S/m)
10	160	0.625
40	97	0.693
100	72	0.89
200	56.5	1.28
300	54	1.37
433	53	1.43
915	51	1.6

4. Using the parameters found in Problem 2.5.3, do a simulation of a pulse striking a medium of muscle tissue. Calculate the frequency domain results at 50, 100, and 433 MHz.

REFERENCES

1. C. M. Furse, S.P. Mathur, and O. P. Gandhi, Improvements to the finite-difference time-domain method for calculating the radar cross section of a perfectly conducting target, *IEEE Trans. Microwave Theory Tech.*, vol. MTT-38, July 1990, pp. 919–927.
2. D. M. Sullivan, Mathematical methods for treatment planning in deep regional hyperthermia, *IEEE Trans. Microwave Theory Tech.*, vol. MTT-39, May 1991, pp. 862–872.
3. R. Luebbers, F. Hunsberger, K. Kunz, R. Standler, and M. Schneider, A frequency-dependent finite-difference time-domain formulation for dispersive materials, *IEEE Trans. Electromagn. Compat.*, vol. EMC-32, Aug. 1990, pp. 222–227.
4. R. M. Joseph, S.C. Hagness, and A. Taflove, Direct time integration of Maxwell's equations in linear dispersive media with absorption for scattering and propagation of femtosecond electromagnetic pulses, *Opt. Lett.*, vol. 16, Sept. 1991, pp. 1412–1411.
5. O. P. Gandhi, B. Q. Gao, and Y. Y. Chen, A frequency-dependent finite-difference time-domain formulation for general dispersive media, *IEEE Trans. Microwave Theory Tech.*, vol. MTT-41, April 1993, pp. 658–665.
6. D. M. Sullivan, Frequency-dependent FDTD methods using Z transforms, *IEEE Trans. Antennas Propag.*, vol. AP-40, Oct. 1992, pp. 1223–1230.
7. A. Ishimaru, *Electromagnetic Wave Propagation, Radiation, and Scattering*, Englewood Cliffs, NJ: Prentice Hall, 1991.
8. M. A. Stuchly and S. S. Stuchly, Dielectric properties of biological substances—tabulated, *J. Microwave Power*, vol. 15, 1980, pp. 16–19.
9. D. M. Sullivan, Use of the finite-difference time-domain method in calculating EM absorption in human tissues, *IEEE Trans. Biomed. Eng.*, vol. BME-34, Feb. 1987, pp. 148–157.
10. D. M. Sullivan, Three dimensional computer simulation in deep regional hyperthermia using the finite-difference time-domain method, *IEEE Trans. Microwave Theory Tech.*, vol. MTT-38, Feb. 1990, pp. 204–211.
11. C. C. Johnson and A. W. Guy, Nonionizing electromagnetic wave effects in biological materials and systems, *Proc. IEEE*, vol. 60, June 1972, pp. 692–718.

PYTHON PROGRAMS USED TO GENERATE FIGURES IN THIS CHAPTER

```
""" fd1d_2_1.py: 1D FDTD

Simulation of a dielectric slab
"""
```

```
import numpy as np
from math import pi, sin
from matplotlib import pyplot as plt

ke = 200
ex = np.zeros(ke)
dx = np.zeros(ke)
ix = np.zeros(ke)
hy = np.zeros(ke)

ddx = 0.01  # Cell size
dt = ddx / 6e8  # Time step size
freq_in = 700e6

boundary_low = [0, 0]
boundary_high = [0, 0]

# Create Dielectric Profile
epsz = 8.854e-12
epsr = 4
sigma = 0.04
k_start = 100

gax = np.ones(ke)
gbx = np.zeros(ke)
gax[k_start:] = 1 / (epsr + (sigma * dt / epsz))
gbx[k_start:] = sigma * dt / epsz

nsteps = 500

# Main FDTD Loop
for time_step in range(1, nsteps + 1):

    # Calculate Dx
    for k in range(1, ke):
        dx[k] = dx[k] + 0.5 * (hy[k - 1] - hy[k])

    # Put a sinusoidal at the low end
    pulse = sin(2 * pi * freq_in * dt * time_step)
    dx[5] = pulse + dx[5]

    # Calculate the Ex field from Dx
    for k in range(1, ke):
        ex[k] = gax[k] * (dx[k] - ix[k])
        ix[k] = ix[k] + gbx[k] * ex[k]

    # Absorbing Boundary Conditions
    ex[0] = boundary_low.pop(0)
    boundary_low.append(ex[1])
```

```
    ex[ke - 1] = boundary_high.pop(0)
    boundary_high.append(ex[ke - 2])

    # Calculate the Hy field
    for k in range(ke - 1):
        hy[k] = hy[k] + 0.5 * (ex[k] - ex[k + 1])

# Plot Fig. 1.6 (generated by method described in Sec. 2.1)
plt.rcParams['font.size'] = 12
plt.figure(figsize=(8, 2.25))

plt.plot(ex, color='k', linewidth=1)
plt.ylabel('E$_x$', fontsize='14')
plt.xticks(np.arange(0, 199, step=20))
plt.xlim(0, 199)
plt.yticks(np.arange(-1, 1.2, step=1))
plt.ylim(-1.2, 1.2)
plt.text(50, 0.5, 'T = {}'.format(time_step),
         horizontalalignment='center')
plt.plot(gbx / gbx[k_start], 'k--',
         linewidth=0.75)  # Scaled for plotting
plt.text(170, 0.5, 'Eps = {}'.format(epsr),
         horizontalalignment='center')
plt.text(170, -0.5, 'Cond = {}'.format(sigma),
         horizontalalignment='center')
plt.xlabel('FDTD cells')

plt.subplots_adjust(bottom=0.25, hspace=0.45)
plt.show()
```

```
""" fd1d_2_2.py: 1D FDTD

The Fourier Transform has been added
"""

import numpy as np
from matplotlib import pyplot as plt
from math import exp, cos, sin, sqrt, atan2

ke = 200
ex = np.zeros(ke)
dx = np.zeros(ke)
ix = np.zeros(ke)
hy = np.zeros(ke)

ddx = 0.01  # Cell size
dt = ddx / 6e8  # Time step size
```

```
number_of_frequencies = 3
freq = np.array((500e6, 200e6, 100e6))

t0 = 50
spread = 10

boundary_low = [0, 0]
boundary_high = [0, 0]

# Create Dielectric Profile
epsz = 8.854e-12
epsr = 4
sigma = 0
k_start = 100

gax = np.ones(ke)
gbx = np.zeros(ke)
gax[k_start:] = 1 / (epsr + (sigma * dt / epsz))
gbx[k_start:] = sigma * dt / epsz

# To be used in the Fourier transform
arg = 2 * np.pi * freq * dt
real_pt = np.zeros((number_of_frequencies, ke))
imag_pt = np.zeros((number_of_frequencies, ke))
real_in = np.zeros(number_of_frequencies)
imag_in = np.zeros(number_of_frequencies)
amp_in = np.zeros(number_of_frequencies)
phase_in = np.zeros(number_of_frequencies)
amp = np.zeros((number_of_frequencies, ke))
phase = np.zeros((number_of_frequencies, ke))

nsteps = 400

# Dictionary to keep track of desired points for plotting
plotting_points = [
    {'num_steps': 200, 'ex': None, 'amp': None, 'phase': None,
     'label': '', 'label_ab': '(a)'},
    {'num_steps': 400, 'ex': None, 'amp': None, 'phase': None,
     'label': 'FDTD cells', 'label_ab': '(b)'}
]

# Main FDTD Loop
for time_step in range(1, nsteps + 1):

    # Calculate Dx
    for k in range(1, ke):
        dx[k] = dx[k] + 0.5 * (hy[k - 1] - hy[k])
```

```
# Put a sinusoidal at the low end
pulse = exp(-0.5 * ((t0 - time_step) / spread) ** 2)
dx[5] = pulse + dx[5]

# Calculate the Ex field from Dx
for k in range(1, ke):
    ex[k] = gax[k] * (dx[k] - ix[k])
    ix[k] = ix[k] + gbx[k] * ex[k]

# Calculate the Fourier transform of Ex
for k in range(ke):
    for m in range(number_of_frequencies):
        real_pt[m, k] = real_pt[m, k] + cos(arg[m]
                        * time_step) \
                        * ex[k]
        imag_pt[m, k] = imag_pt[m, k] - sin(arg[m]
                        * time_step) \
                        * ex[k]

# Fourier Transform of the input pulse
if time_step < 100:
    for m in range(number_of_frequencies):
        real_in[m] = real_in[m] + cos(arg[m] * time_step)
                     * ex[10]
        imag_in[m] = imag_in[m] - sin(arg[m] * time_step)
                     * ex[10]

# Absorbing Boundary Conditions
ex[0] = boundary_low.pop(0)
boundary_low.append(ex[1])

ex[ke - 1] = boundary_high.pop(0)
boundary_high.append(ex[ke - 2])

# Calculate the Hy field
for k in range(ke - 1):
    hy[k] = hy[k] + 0.5 * (ex[k] - ex[k + 1])

# Save data at certain points for later plotting
for plotting_point in plotting_points:
    if time_step == plotting_point['num_steps']:

        # Calculate the amplitude and phase at each freq
        for m in range(number_of_frequencies):
            amp_in[m] = sqrt(imag_in[m] ** 2 + real_in [m]
                        ** 2)
            phase_in[m] = atan2(imag_in[m], real_in[m])
```

```
            for k in range(ke):
                amp[m, k] = (1 / amp_in[m]) * sqrt(
                    (real_pt[m, k]) ** 2 + (imag_pt[m, k])
                       ** 2)
                phase[m, k] = atan2(imag_pt[m, k],
                                    real_pt[m, k]) \
                                    - phase_in[m]

        plotting_point['ex'] = np.copy(ex)
        plotting_point['amp'] = np.copy(amp)
        plotting_point['phase'] = np.copy(phase)

# Plot the outputs in Fig. 2.1
plt.rcParams['font.size'] = 12
fig = plt.figure(figsize=(8, 7))


def plot_e_field(data, ga, timestep, label_ab):
    """Plot of E field at a single time step"""

    plt.plot(data, color='k', linewidth=1)
    plt.ylabel('E$_x$', fontsize='14')
    plt.xticks(np.arange(0, 199, step=20))
    plt.xlim(0, 199)
    plt.yticks((-1, 0, 1))
    plt.ylim(-1.2, 1.2)
    plt.text(35, 0.3, 'Time Domain, T = {}'.format(timestep),
             horizontalalignment='center')
    plt.plot(-(ga - 1) / 0.75, 'k--',
             linewidth=0.75) # The math on gb is just for scaling
    plt.text(-25, -2.1, label_ab, horizontalalignment='center')
    return


def plot_amp(data, ga, freq, label):
    """ Plot of the Fourier transform amplitude at a single time
        step"""

    plt.plot(data[0], color='k', linewidth=1)
    plt.ylabel('Amp')
    plt.xticks(np.arange(0, 199, step=20))
    plt.xlim(0, 199)
    plt.yticks(np.arange(0, 1.9, step=1))
    plt.ylim(0, 1.9)
    plt.text(150, 1.2,
             'Freq. Domain at {} MHz'.format(int(round
               (freq[0] / 1e6))),
             horizontalalignment='center')
```

```
    plt.plot(-(ga - 1) / 0.75, 'k--',
             linewidth=0.75) # The math on gb is just for scaling
    plt.xlabel('{}'.format(label))
    return


# Plot the E field at each of the time steps saved earlier
for subplot_num, plotting_point in enumerate(plotting_points):
    ax = fig.add_subplot(4, 1, subplot_num * 2 + 1)
    plot_e_field(plotting_point['ex'], gax,
                 plotting_point['num_steps'],
                 plotting_point['label_ab'])
    ax = fig.add_subplot(4, 1, subplot_num * 2 + 2)
    plot_amp(plotting_point['amp'], gax, freq,
             plotting_point['label'], )

plt.subplots_adjust(bottom=0.1, hspace=0.45)
plt.show()

""" fd1d_2_3.py: 1D FDTD

Simulation of a frequency-dependent material
"""

import numpy as np
from matplotlib import pyplot as plt
from math import pi, exp, cos, sin, sqrt, atan2

ke = 200
ex = np.zeros(ke)
dx = np.zeros(ke)
ix = np.zeros(ke)
sx = np.zeros(ke)
hy = np.zeros(ke)

ddx = 0.01  # Cell size
dt = ddx / 6e8 # Time step size
number_of_frequencies = 3
freq_in = np.array((50e6, 200e6, 500e6))

t0 = 50
spread = 10

boundary_low = [0, 0]
boundary_high = [0, 0]
```

```
# Create Dielectric Profile
epsz = 8.854e-12
epsr = 2
sigma = 0.01
tau = 0.001 * 1e-6
chi = 2
k_start = 100

gax = np.ones(ke)
gbx = np.zeros(ke)
gcx = np.zeros(ke)
gax[k_start:] = 1 / (epsr + (sigma * dt / epsz) + chi * dt / tau)
gbx[k_start:] = sigma * dt / epsz
gcx[k_start:] = chi * dt / tau
del_exp = exp(-dt / tau)

# To be used in the Fourier transform
arg = 2 * np.pi * freq_in * dt
real_pt = np.zeros((number_of_frequencies, ke))
imag_pt = np.zeros((number_of_frequencies, ke))
real_in = np.zeros(number_of_frequencies)
imag_in = np.zeros(number_of_frequencies)
amp_in = np.zeros(number_of_frequencies)
phase_in = np.zeros(number_of_frequencies)
amp = np.zeros((number_of_frequencies, ke))
phase = np.zeros((number_of_frequencies, ke))

nsteps = 1000

# Dictionary to keep track of desired points for plotting
plotting_points = [
    {'num_steps': 250, 'ex': None, 'scaling_factor': 1,
     'gb_scaling_factor': 1, 'y_ticks': (0, 0.5, 1),
     'y_min': -0.3, 'y_max': 1.2, 'y_text_loc': 0.3,
     'label': '(a)', 'label_loc': 0.4},
    {'num_steps': 1000, 'ex': None, 'scaling_factor': 1000,
     'gb_scaling_factor': 2, 'y_ticks': (-6, -4, -2, 0, 2),
     'y_min': -6.2, 'y_max': 2.2, 'y_text_loc': -3,
     'label': '(b)', 'label_loc': -1.9}
]

# Main FDTD Loop
for time_step in range(1, nsteps + 1):

    # Calculate Dx
    for k in range(1, ke):
        dx[k] = dx[k] + 0.5 * (hy[k - 1] - hy[k])
```

```
    # Put a sinusoidal at the low end
    pulse = exp(-0.5 * ((t0 - time_step) / spread) ** 2)
    dx[5] = pulse + dx[5]

    # Calculate the Ex field from Dx
    for k in range(1, ke):
        ex[k] = gax[k] * (dx[k] - ix[k] - del_exp * sx[k])
        ix[k] = ix[k] + gbx[k] * ex[k]
        sx[k] = del_exp * sx[k] + gcx[k] * ex[k]

    # Calculate the Fourier transform of Ex
    for k in range(ke):
        for m in range(number_of_frequencies):
            real_pt[m, k] = real_pt[m, k] + cos(arg[m] *
                            time_step) \
                            * ex[k]
            imag_pt[m, k] = imag_pt[m, k] - sin(arg[m] *
                            time_step) \
                            * ex[k]

    # Fourier Transform of the input pulse
    if time_step < 3 * t0:
        for m in range(number_of_frequencies):
            real_in[m] = real_in[m] + cos(arg[m] * time_step)
                         * ex[10]
            imag_in[m] = imag_in[m] - sin(arg[m] * time_step)
                         * ex[10]

    # Absorbing Boundary Conditions
    ex[0] = boundary_low.pop(0)
    boundary_low.append(ex[1])

    ex[ke - 1] = boundary_high.pop(0)
    boundary_high.append(ex[ke - 2])

    # Calculate the Hy field
    for k in range(ke - 1):
        hy[k] = hy[k] + 0.5 * (ex[k] - ex[k + 1])

    # Save data at certain points for later plotting
    for plotting_point in plotting_points:
        if time_step == plotting_point['num_steps']:
            plotting_point['ex'] = np.copy(ex)

# Calculate the amplitude and phase at each frequency
for m in range(number_of_frequencies):
    amp_in[m] = sqrt(imag_in[m] ** 2 + real_in[m] ** 2)
    phase_in[m] = atan2(imag_in[m], real_in[m])
```

```
    for k in range(ke):
        amp[m, k] = (1 / amp_in[m]) * sqrt(
            (real_pt[m, k]) ** 2 + (imag_pt[m, k]) ** 2)
        phase[m, k] = atan2(imag_pt[m, k], real_pt[m, k])
          - phase_in[m]

# Plot the outputs in Fig. 2.3
plt.rcParams['font.size'] = 12
fig = plt.figure(figsize=(8, 7))


def plot_e_field(data, gb, timestep, scaling_factor,
  gb_scaling_factor,
                 y_ticks, y_min, y_max, y_text_loc, label,
                   label_loc):
    """Plot of E field at a single time step"""

    plt.plot(data * scaling_factor, color='k', linewidth=1)
    plt.ylabel('E$_x$ (V/m)', fontsize='12')
    plt.xticks(np.arange(0, 199, step=20))
    plt.xlim(0, 199)
    plt.yticks(y_ticks)
    plt.ylim(y_min, y_max)
    plt.text(150, y_text_loc, 'Time Domain, T = {}'.
      format(timestep),
             horizontalalignment='center')
    plt.plot(gb * gb_scaling_factor / gb[120], 'k--',
             linewidth=0.75) # The math on gb is just for scaling
    plt.text(-25, label_loc, label, horizontalalignment=
      'center')

    return


# Plot the E field at each of the time steps saved earlier
for subplot_num, plotting_point in enumerate(plotting_points):
    ax = fig.add_subplot(5, 1, subplot_num + 1)
    plot_e_field(plotting_point['ex'], gbx,
                 plotting_point['num_steps'],
                 plotting_point['scaling_factor'],
                 plotting_point['gb_scaling_factor'],
                 plotting_point['y_ticks'], plotting_
                   point['y_min'],
                 plotting_point['y_max'],
                 plotting_point['y_text_loc'],
                   plotting_point['label'],
                 plotting_point['label_loc'])
```

```
# Dictionary to keep track of plotting for the amplitudes
plotting_freqs = [
    {'freq': freq_in[0], 'amp': amp[0], 'label': '(c)',
     'x_label': ''},
    {'freq': freq_in[1], 'amp': amp[1], 'label': '',
     'x_label': ''},
    {'freq': freq_in[2], 'amp': amp[2], 'label': '',
     'x_label': 'FDTD cells'}
]


def plot_amp(data, gb, freq, label, x_label):
    """Plot of amplitude at one frequency"""
    plt.plot(data, color='k', linewidth=1)
    plt.ylabel('Amp')
    plt.xticks(np.arange(0, 199, step=20))
    plt.xlim(0, 198)
    plt.yticks(np.arange(0, 2.1, step=1))
    plt.ylim(-0.2, 2.0)
    plt.text(150, 1.2, 'Freq. Domain at {} MHz'.format
      (int(round(
        freq / 1e6))), horizontalalignment='center')
    plt.plot(gb * 1 / gb[120], 'k--',
             linewidth=0.75) # The math on gb is just for scaling
    plt.text(-25, 0.6, label, horizontalalignment='center')
    plt.xlabel(x_label)

    return


# Plot the amplitude at each of the frequencies of interest
for subplot_num, plotting_freq in enumerate(plotting_freqs):
    ax = fig.add_subplot(5, 1, subplot_num + 3)
    plot_amp(plotting_freq['amp'], gbx, plotting_freq['freq'],
             plotting_freq['label'], plotting_freq['x_label'])

plt.subplots_adjust(bottom=0.1, hspace=0.45)
plt.show()
```

3

TWO-DIMENSIONAL SIMULATION

This chapter introduces two-dimensional simulation. We begin with the basic two-dimensional formulation in FDTD and a simple example using a point source. Then, the absorbing boundary conditions (ABCs) are described, along with their implementation into the FDTD program. Finally, the generation of electromagnetic plane waves using FDTD is described.

3.1 FDTD IN TWO DIMENSIONS

Once again, we will start with the normalized Maxwell's equations that we used in Chapter 2:

$$\frac{\partial \widetilde{\boldsymbol{D}}}{\partial t} = \frac{1}{\sqrt{\varepsilon_0 \mu_0}} \nabla \times \boldsymbol{H}, \tag{3.1a}$$

$$\widetilde{\boldsymbol{D}}(\omega) = \varepsilon_r^*(\omega) \cdot \widetilde{\boldsymbol{E}}(\omega), \tag{3.1b}$$

$$\frac{\partial \boldsymbol{H}}{\partial t} = -\frac{1}{\sqrt{\varepsilon_0 \mu_0}} \nabla \times \widetilde{\boldsymbol{E}}. \tag{3.1c}$$

Electromagnetic Simulation Using the FDTD Method with Python, Third Edition.
Jennifer E. Houle and Dennis M. Sullivan.

Published 2020 by John Wiley & Sons, Inc.

When we get to three-dimensional simulation, we will wind up dealing with six different fields: $\widetilde{E}_x$, $\widetilde{E}_y$, $\widetilde{E}_z$, H_x, H_y, and H_z. For two-dimensional simulation, we choose between one of the two groups of three vectors each: (i) the transverse magnetic (TM) mode, which is composed of $\widetilde{E}_z$, H_x, and H_y, or (ii) the transverse electric (TE) mode, which is composed of $\widetilde{E}_x$, $\widetilde{E}_y$, and H_z. We will work with the TM mode. Equation (3.1) is now reduced to

$$\frac{\partial D_z}{\partial t} = \frac{1}{\sqrt{\varepsilon_0 \mu_0}} \left(\frac{\partial H_y}{\partial x} - \frac{\partial H_x}{\partial y} \right), \tag{3.2a}$$

$$D_z(\omega) = \varepsilon_r^*(\omega) \cdot E_z(\omega), \tag{3.2b}$$

$$\frac{\partial H_x}{\partial t} = -\frac{1}{\sqrt{\varepsilon_0 \mu_0}} \frac{\partial E_z}{\partial y}, \tag{3.2c}$$

$$\frac{\partial H_y}{\partial t} = \frac{1}{\sqrt{\varepsilon_0 \mu_0}} \frac{\partial E_z}{\partial x}. \tag{3.2d}$$

(Once again, we have dropped the tildes over the E and D fields, but it is understood that we are using the normalized units described in Chapter 1.)

As in one-dimensional simulation, it is important that there is a systematic interleaving of the fields to be calculated. This is illustrated in Fig. 3.1. Putting Eq. (3.2a), (3.2c), and (3.2d) into the finite-differencing scheme results in the following difference equations (1):

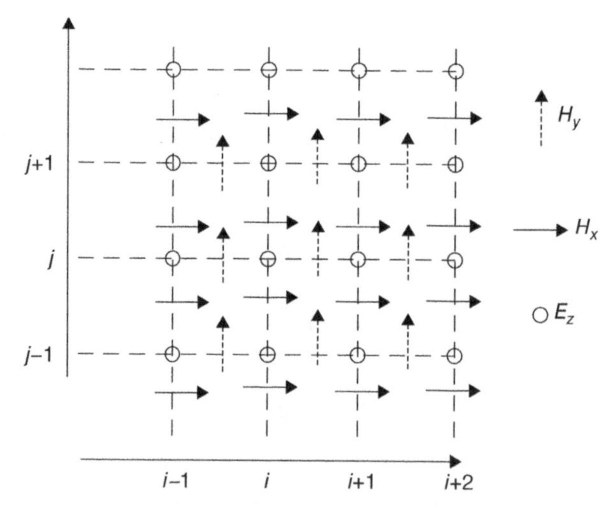

Figure 3.1 Interleaving of the E and H fields for the two-dimensional TM formulation.

$$\frac{D_z^{n+1/2}(i,j) - D_z^{n-1/2}(i,j)}{\Delta t} = \frac{1}{\sqrt{\varepsilon_0 \mu_0}} \left[\frac{H_y^n\left(i+\frac{1}{2},j\right) - H_y^n\left(i-\frac{1}{2},j\right)}{\Delta x} \right] - \frac{1}{\sqrt{\varepsilon_0 \mu_0}} \left[\frac{H_x^n\left(i,j+\frac{1}{2}\right) - H_x^n\left(i,j-\frac{1}{2}\right)}{\Delta x} \right], \tag{3.3a}$$

$$\frac{H_x^{n+1}\left(i,j+\frac{1}{2}\right) - H_x^n\left(i,j+\frac{1}{2}\right)}{\Delta t} = -\frac{1}{\sqrt{\varepsilon_0 \mu_0}} \frac{E_z^{n+1/2}(i,j+1) - E_z^{n+1/2}(i,j)}{\Delta x}, \tag{3.3b}$$

$$\frac{H_y^{n+1}\left(i+\frac{1}{2},j\right) - H_y^n\left(i+\frac{1}{2},j\right)}{\Delta t} = \frac{1}{\sqrt{\varepsilon_0 \mu_0}} \frac{E_z^{n+1/2}(i+1,j) - E_z^{n+1/2}(i,j)}{\Delta x}. \tag{3.3c}$$

For this book, we assume each cell will have equal dimensions, so $\Delta x = \Delta y$ in two-dimensional simulation. This allows Δx to be used exclusively and simplifies the equations. Using the same type of manipulation as in Chapter 1, including

$$\Delta t = \frac{\Delta x}{2 \cdot c_0},$$

where c_0 is the speed of light in free space, we get the equations:

```
dz[i, j] = dz[i, j] + 0.5 * (hy[i, j] - hy[i - 1, j] - hx[i, j]
           + hx[i, j - 1])                                        (3.4a)
ez[i, j] = gaz[i, j] * dz[i, j]                                   (3.4b)
hx[i, j] = hx[i, j] + 0.5 * (ez[i, j] - ez[i, j + 1])             (3.4c)
hy[i, j] = hy[i, j] + 0.5 * (ez[i + 1, j] - ez[i, j])             (3.4d)
```

Note that the relationship between E_z and D_z is the same as that for the simple lossy dielectric in the one-dimensional case. Obviously, the same modification can be made to include frequency-dependent terms.

The program fd2d_3_1.py implements the above equations. It has a simple Gaussian pulse source that is generated in the middle of the problem space. Figure 3.2 presents the results of a simulation for the first 50 time steps.

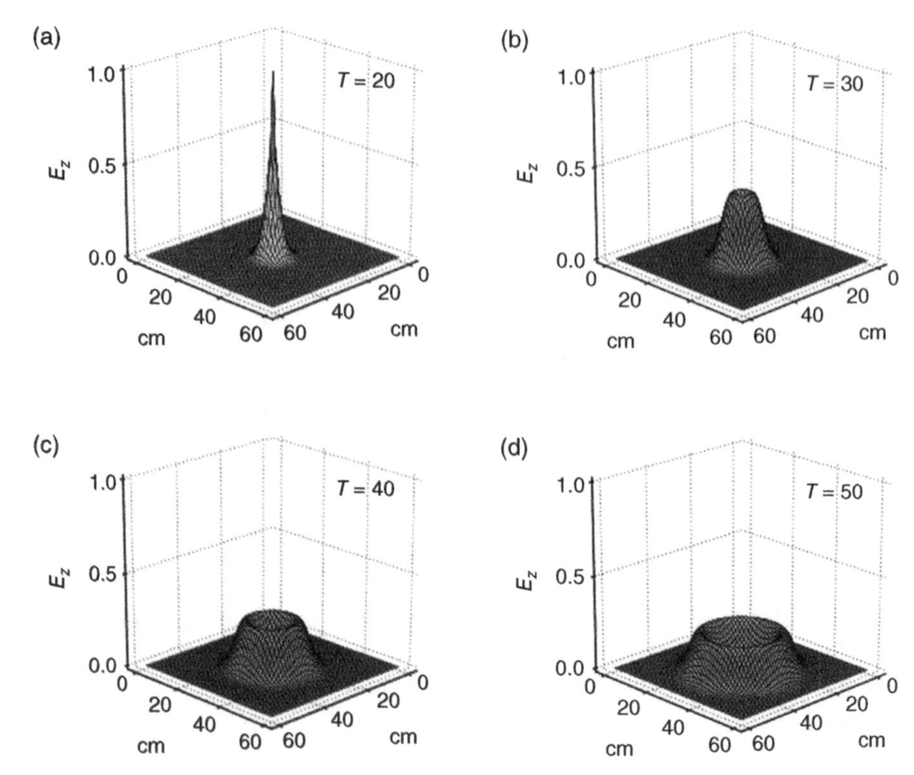

Figure 3.2 Results of a simulation using the program fd2d_3_1.py. A Gaussian pulse is initiated in the middle and travels outward. (a) $T=20$, (b) $T=30$, (c) $T=40$, and (d) $T=50$.

PROBLEM SET 3.1

1. Get the program fd2d_3_1.py running. Duplicate the results of Fig. 3.2. Let it run until it hits the boundary. What happens?

3.2 THE PERFECTLY MATCHED LAYER (PML)

Until now, we have only briefly mentioned the issue of ABCs. The size of the area that can be simulated using FDTD is limited by computer resources. For instance, in the two-dimensional simulation of the previous section, the program contains two-dimensional matrices for the values of all the fields, `dz`, `ez`, `hx`, and `hy`, as well as a matrix to hold the parameter `gaz`. (Note the `ga` parameter used in this chapter is in the z direction, as shown in Eq. (3.4b).)

Suppose we are simulating a wave generated from a point source propagating in free space as in Fig. 3.2. As the wave propagates outward, it will eventually come to the edge of the allowable space, which is dictated by how the matrices have been dimensioned in the program. If nothing were done to address this, reflections would be generated that propagate inward. There would be no way to determine the real wave from the reflected junk. This is the reason that ABCs have been an issue for as long as FDTD has been used. There have been numerous approaches to this problem (2, 3).

The most flexible and efficient ABC is the perfectly matched layer (PML) developed by Berenger (4). The basic idea behind the PML is that if a wave is propagating in medium A and it impinges on medium B, the amount of reflection is dictated by the intrinsic impedances of the two media as given by the equation

$$\Gamma = \frac{\eta_B - \eta_A}{\eta_B + \eta_A}, \tag{3.5}$$

where Γ is the reflection coefficient, and η_A and η_B are the impedances of the media as calculated from the dielectric constants ε and the permeabilities μ of the two media:

$$\eta = \sqrt{\frac{\mu}{\varepsilon_r \varepsilon_0}}. \tag{3.6}$$

So far, we have assumed that μ was a constant, so when a propagating pulse went from $\varepsilon_r = 1$ to $\varepsilon_r = 4$ as in Fig. 2.1, it saw a change in impedance and reflected a portion of the pulse given by Eq. (3.5). However, if μ changed with ε so that η remained a constant, Γ would remain constant and no reflection would occur. This still does not solve our problem because the pulse will continue propagating in the new medium. What we really want is a medium that is also lossy so the pulse will die out before it reaches the end of the problem space. This is accomplished by making both ε and μ complex, because the imaginary part represents the part that causes decay (Appendix 1.A).

Let us return to Eq. (3.2) but move everything to the Fourier domain. (We are going to the Fourier domain in *time*, so d/dt becomes $j\omega$. This does not affect the spatial derivatives.)

$$j\omega D_z = c_0 \cdot \left(\frac{\partial H_y}{\partial x} - \frac{\partial H_x}{\partial y} \right), \tag{3.7a}$$

$$D_z(\omega) = \varepsilon_r^*(\omega) \cdot E_z(\omega), \tag{3.7b}$$

$$j\omega H_x = -c_0 \frac{\partial E_z}{\partial y}, \tag{3.7c}$$

$$j\omega H_y = c_0 \frac{\partial E_z}{\partial x}. \tag{3.7d}$$

Remember that we have eliminated ε and μ from the spatial derivatives in Eq. (3.7a), (3.7c), and (3.7d) for the normalized units, and $c_0 = 1/\sqrt{\varepsilon_0\mu_0}$. Instead of putting them back to implement the PML, we will add the fictitious dielectric constant and permeabilities ε^*_{Fz}, μ^*_{Fx}, and μ^*_{Fy} (5):

$$j\omega D_z \cdot \varepsilon^*_{Fz}(x) \cdot \varepsilon^*_{Fz}(y) = c_0 \cdot \left(\frac{\partial H_y}{\partial x} - \frac{\partial H_x}{\partial y}\right), \tag{3.8a}$$

$$D_z(\omega) = \varepsilon^*_r(\omega) \cdot E_z(\omega), \tag{3.8b}$$

$$j\omega H_x \cdot \mu^*_{Fx}(x) \cdot \mu^*_{Fx}(y) = -c_0 \frac{\partial E_z}{\partial y}, \tag{3.8c}$$

$$j\omega H_y \cdot \mu^*_{Fy}(x) \cdot \mu^*_{Fy}(y) = c_0 \frac{\partial E_z}{\partial x}. \tag{3.8d}$$

Two things are worth noting: first, the value ε^*_{Fz} is associated with the flux density D, not the electric field E; and second, we have added two values each of ε^*_{Fz} in Eq. (3.8a), and μ^*_{Fx} and μ^*_{Fy} in Eq. (3.8c) and (3.8d), respectively. These fictitious values to implement the PML have nothing to do with the *real* values of $\varepsilon^*_r(\omega)$, which specify the medium.

Sacks et al. (6) showed that there are two conditions to form a PML:

1. The impedance going from the background medium to the PML must be constant,

$$\eta_0 = \eta_m = \sqrt{\frac{\mu^*_{Fx}}{\varepsilon^*_{Fx}}} = 1. \tag{3.9}$$

   The impedance is 1 because of our normalized units.

2. In the direction perpendicular to the boundary (e.g., the x direction), the relative dielectric constant and relative permeability must be the inverse of those in the other directions, that is,

$$\varepsilon^*_{Fx} = \frac{1}{\varepsilon^*_{Fy}}, \tag{3.10a}$$

$$\mu^*_{Fx} = \frac{1}{\mu^*_{Fy}}. \tag{3.10b}$$

We will assume that each of these is a complex quantity of the form:

$$\varepsilon^*_{Fm} = \varepsilon_{Fm} + \frac{\sigma_{Dm}}{j\omega\varepsilon_0} \text{ for } m = x \text{ or } y, \tag{3.11a}$$

$$\mu_{Fm}^{*} = \mu_{Fm} + \frac{\sigma_{Dm}}{j\omega\mu_0} \text{ for } m = x \text{ or } y. \tag{3.11b}$$

The following selection of parameters satisfies Eq. (3.10a) and (3.10b) (7):

$$\varepsilon_{Fm} = \mu_{Fm} = 1, \tag{3.12a}$$

$$\frac{\sigma_{Dm}}{\varepsilon_0} = \frac{\sigma_{Dm}}{\mu_0} = \frac{\sigma_D}{\varepsilon_0}. \tag{3.12b}$$

Substituting Eq. (3.12) into (3.11), the value in Eq. (3.9) becomes

$$\eta_0 = \eta_m = \sqrt{\frac{\mu_{Fx}^{*}}{\varepsilon_{Fx}^{*}}} = \sqrt{\frac{1 + \dfrac{\sigma_D(x)}{j\omega\varepsilon_0}}{1 + \dfrac{\sigma_D(x)}{j\omega\varepsilon_0}}} = 1.$$

This fulfills the first requirement mentioned earlier. If σ_D increases gradually as it goes into the PML, Eq. (3.8a), (3.8c), and (3.8d) will cause D_z, H_x, and H_y to be attenuated.

We will start by implementing a PML only in the X direction and retain only the x dependent values of ε_F^{*} and μ_F^{*} in Eq. (3.8):

$$j\omega D_z \cdot \varepsilon_{Fz}^{*}(x) = c_0 \cdot \left(\frac{\partial H_y}{\partial x} - \frac{\partial H_x}{\partial y} \right),$$

$$j\omega H_x \cdot \mu_{Fx}^{*}(x) = -c_0 \frac{\partial E_z}{\partial y},$$

$$j\omega H_y \cdot \mu_{Fy}^{*}(x) = c_0 \frac{\partial E_z}{\partial x},$$

and use the values of Eq. (3.12):

$$j\omega \left[1 + \frac{\sigma_D(x)}{j\omega\varepsilon_0} \right] D_z = c_0 \cdot \left(\frac{\partial H_y}{\partial x} - \frac{\partial H_x}{\partial y} \right), \tag{3.13a}$$

$$j\omega \left[1 + \frac{\sigma_D(x)}{j\omega\varepsilon_0} \right]^{-1} H_x = -c_0 \frac{\partial E_z}{\partial y}, \tag{3.13b}$$

$$j\omega \left[1 + \frac{\sigma_D(x)}{j\omega\varepsilon_0} \right] H_y = c_0 \frac{\partial E_z}{\partial x}. \tag{3.13c}$$

Note that the permeability of H_x in Eq. (3.13b) is the inverse of that of H_y in Eq. (3.13c) in keeping with Eq. (3.10b). In this way, we have fulfilled the second requirement for the PML. (Eq. (3.10a) is irrelevant for this two-dimensional case because we only have an E field in the z direction, which is perpendicular to both x and y, the directions of propagation.)

Now, Eq. (3.13) must be entered into the FDTD formulation. First, consider the left side of Eq. (3.13a):

$$j\omega\left[1+\frac{\sigma_D(x)}{j\omega\varepsilon_0}\right]D_z = j\omega D_z + \frac{\sigma_D(x)}{\varepsilon_0}D_z.$$

Moving to the time domain and taking the finite-difference approximations, we get the following:

$$\begin{aligned} j\omega D_z + \frac{\sigma_D(x)}{\varepsilon_0}D_z &\cong \frac{D_z^{n+1/2}(i,j)-D_z^{n-1/2}(i,j)}{\Delta t} + \left[\frac{\sigma_D(i)}{\varepsilon_0}\right]\left[\frac{D_z^{n+1/2}(i,j)+D_z^{n-1/2}(i,j)}{2}\right] \\ &= D_z^{n+1/2}(i,j)\cdot\frac{1}{\Delta t}\left[1+\frac{\sigma_D(i)\cdot\Delta t}{2\cdot\varepsilon_0}\right] - D_z^{n-1/2}(i,j)\cdot\frac{1}{\Delta t}\left[1-\frac{\sigma_D(i)\cdot\Delta t}{2\cdot\varepsilon_0}\right]. \end{aligned}$$

Notice that D_z is averaged over two time steps in the second term because the derivative uses two time steps. If we put this into Eq. (3.13a), along with the spatial derivatives, we get

$$\begin{aligned} D_z^{n+1/2}(i,j) &= gi3(i)\cdot D_z^{n-1/2}(i,j) \\ &+ gi2(i)\cdot 0.5\cdot\left[H_y^n\left(i+\frac{1}{2},j\right)-H_y^n\left(i-\frac{1}{2},j\right)-H_x^n\left(i,j+\frac{1}{2}\right)+H_x^n\left(i,j-\frac{1}{2}\right)\right]. \end{aligned} \tag{3.14}$$

Once again, from Eq. (1.8), we have used the fact that

$$\frac{\Delta t}{\Delta x}c_0 = \frac{\frac{\Delta x}{2c_0}}{\Delta x}c_0 = \frac{1}{2}.$$

The new parameters $gi2$ and $gi3$ are given by

$$gi2(i) = \frac{1}{1+\sigma_D(i)\cdot\dfrac{\Delta t}{2\cdot\varepsilon_0}}, \tag{3.15a}$$

$$gi3(i) = \frac{1-\sigma_D(i)\cdot\dfrac{\Delta t}{2\cdot\varepsilon_0}}{1+\sigma_D(i)\cdot\dfrac{\Delta t}{2\cdot\varepsilon_0}}. \tag{3.15b}$$

An almost identical treatment of Eq. (3.13c) gives

$$H_y^{n+1}\left(i+\frac{1}{2},j\right)=fi3\left(i+\frac{1}{2}\right)\cdot H_y^n\left(i+\frac{1}{2},j\right) \\ +fi2\left(i+\frac{1}{2}\right)\cdot 0.5\cdot\left[E_z^{n+1/2}(i+1,j)-E_z^{n+1/2}(i,j)\right], \tag{3.16}$$

where

$$fi2\left(i+\frac{1}{2}\right)=\frac{1}{1+\sigma_D\left(i+\frac{1}{2}\right)\cdot\dfrac{\Delta t}{2\cdot\varepsilon_0}}, \tag{3.17a}$$

$$fi3\left(i+\frac{1}{2}\right)=\frac{1-\sigma_D\left(i+\frac{1}{2}\right)\cdot\dfrac{\Delta t}{2\cdot\varepsilon_0}}{1+\sigma_D\left(i+\frac{1}{2}\right)\cdot\dfrac{\Delta t}{2\cdot\varepsilon_0}}. \tag{3.17b}$$

Notice that these parameters are calculated at $i + 1/2$ because of the position of H_y in the FDTD grid (Fig. 3.1).

Equation (3.13b) will require a somewhat different treatment than the previous two equations. Start by rewriting it as

$$j\omega H_x=-c_0\left[\frac{\partial E_z}{\partial y}+\frac{\sigma_D(x)}{j\omega\varepsilon_0}\frac{\partial E_z}{\partial y}\right].$$

Remember that $1/(j\omega)$ may be regarded as an integration operator in time and $j\omega$ as a derivative in time. The spatial derivative will be written as

$$\frac{\partial E_z}{\partial y}\cong\frac{E_z^{n+1/2}(i,j+1)-E_z^{n+1/2}(i,j)}{\Delta x}=-\frac{curl\_e}{\Delta x}.$$

Note the negative sign in front of *curl_e*. Implementing this into an FDTD formulation gives

$$\frac{H_x^{n+1}\left(i,j+\frac{1}{2}\right)-H_x^n\left(i,j+\frac{1}{2}\right)}{\Delta t}=-c_0\left[-\frac{curl\_e}{\Delta x}-\frac{\sigma_D(i)}{\varepsilon_0}\cdot\Delta t\cdot\sum_{n=0}^{T}\frac{curl\_e}{\Delta x}\right].$$

Note the extra Δt in front of the summation. This is part of the approximation of the time-domain integral. Finally, we get

$$H_x^{n+1}\left(i,j+\frac{1}{2}\right) = H_x^n\left(i,j+\frac{1}{2}\right) + \frac{c_0 \cdot \Delta t}{\Delta x}\left[curl\_e + \frac{\sigma_D(i) \cdot \Delta t}{\varepsilon_0} I_{Hx}^{n+1/2}\left(i,j+\frac{1}{2}\right)\right].$$

The parameter $I_{Hx}^{n+1/2}$ is introduced for the summation. Equation (3.13b) is implemented as the following series of equations:

$$curl\_e = \left[E_z^{n+1/2}(i,j) - E_z^{n+1/2}(i,j+1)\right], \tag{3.18a}$$

$$I_{Hx}^{n+1/2}\left(i,j+\frac{1}{2}\right) = I_{Hx}^{n-1/2}\left(i,j+\frac{1}{2}\right) + curl\_e, \tag{3.18b}$$

$$H_x^{n+1}\left(i,j+\frac{1}{2}\right) = H_x^n\left(i,j+\frac{1}{2}\right) + \left[0.5 \cdot curl\_e + fi1(i) \cdot I_{Hx}^{n+1/2}\left(i,j+\frac{1}{2}\right)\right], \tag{3.18c}$$

with

$$fi1(i) = \frac{\sigma_D(i) \cdot \Delta t}{2\varepsilon_0}. \tag{3.19}$$

In calculating the f and g parameters, it is not necessary to actually vary conductivities. Instead, we calculate an auxiliary parameter,

$$xn(i) = \frac{\sigma_D(i) \cdot \Delta t}{2\varepsilon_0},$$

which increases as it goes into the PML. The f and g parameters are then calculated:

$$xn(i) = 0.333 * \left(\frac{i}{\text{length\_pml}}\right)^3 \quad i = 1, 2, \ldots, \text{length\_pml}, \tag{3.20}$$

$$fi1(i) = xn(i), \tag{3.21a}$$

$$gi2(i) = \left[\frac{1}{1 + xn(i)}\right], \tag{3.21b}$$

$$gi3(i) = \left[\frac{1 - xn(i)}{1 + xn(i)}\right]. \tag{3.21c}$$

Notice that the quantity in parentheses in Eq. (3.20) ranges between 0 and 1. The factor 0.333 was found empirically to be the largest number that remained stable. Similarly, the cubic factor in Eq. (3.20) is found empirically to be the most effective variation. (The parameters $fi2$ and $fi3$ are different only because they are computed at the half intervals $i + 1/2$.) The parameters vary in the following manner:

$$fi1(i) \text{ from } 0 \text{ to } 0.333 \tag{3.22a}$$

$$gi2(i) \text{ from } 1 \text{ to } 0.75 \tag{3.22b}$$

$$gi3(i) \text{ from } 1 \text{ to } 0.5 \tag{3.22c}$$

Throughout the main problem space, $fi1$ is 0, while $gi2$ and $gi3$ are 1. Therefore, there is a "seamless" transition from the main part of the program to the PML (Fig. 3.3).

So far, we have shown the implementation of the PML in the x direction. Obviously, it must also be done in the y direction. To do this, we must go back and add the y-dependent terms from Eq. (3.8) that were set aside. So, instead of Eq. (3.13) we have

$$j\omega\left[1 + \frac{\sigma_D(x)}{j\omega\varepsilon_0}\right]\left[1 + \frac{\sigma_D(y)}{j\omega\varepsilon_0}\right]D_z = c_0\cdot\left(\frac{\partial H_y}{\partial x} - \frac{\partial H_x}{\partial y}\right), \tag{3.23a}$$

$$j\omega\left[1 + \frac{\sigma_D(x)}{j\omega\varepsilon_0}\right]^{-1}\left[1 + \frac{\sigma_D(y)}{j\omega\varepsilon_0}\right]H_x = -c_0\frac{\partial E_z}{\partial y}, \tag{3.23b}$$

$$j\omega\left[1 + \frac{\sigma_D(x)}{j\omega\varepsilon_0}\right]\left[1 + \frac{\sigma_D(y)}{j\omega\varepsilon_0}\right]^{-1}H_y = c_0\frac{\partial E_z}{\partial x}. \tag{3.23c}$$

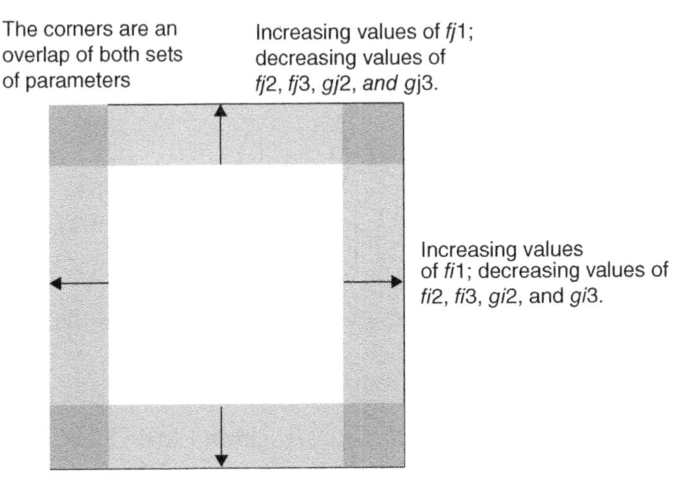

Figure 3.3 Parameters related to the perfectly matched layer (PML).

Using the same procedure as before, the following replaces Eq. (3.14):

$$D_z^{n+1/2}(i,j) = gi3(i)\cdot gj3(j)\cdot D_z^{n-1/2}(i,j) + gi2(i)\cdot gj2(j)\cdot 0.5\cdot \left[H_y^n\left(i+\frac{1}{2},j\right) - H_y^n\left(i-\frac{1}{2},j\right) - H_x^n\left(i,j+\frac{1}{2}\right) + H_x^n\left(i,j-\frac{1}{2}\right)\right].$$

Technically, this term would be more complicated after multiplying out Eq. (3.23a) and requires an integral. However, this is found to be sufficient in absorbing outgoing waves.

In the y direction, H_y will require an implementation similar to the one used for H_x in the x direction giving

$$curl\_e = \left[E_z^{n+1/2}(i,j) - E_z^{n+1/2}(i+1,j)\right], \tag{3.24a}$$

$$I_{Hy}^{n+1/2}\left(i+\frac{1}{2},j\right) = I_{Hy}^{n-1/2}\left(i+\frac{1}{2},j\right) + curl\_e, \tag{3.24b}$$

$$H_y^{n+1}\left(i+\frac{1}{2},j\right) = fi3\left(i+\frac{1}{2}\right)\cdot H_y^n\left(i+\frac{1}{2},j\right) - fi2\left(i+\frac{1}{2}\right)\cdot\left[0.5\cdot curl\_e + fj1(j)\cdot I_{Hy}^{n+1/2}\left(i+\frac{1}{2},j\right)\right]. \tag{3.24c}$$

Finally, the H_x equation becomes

$$curl\_e = \left[E_z^{n+1/2}(i,j) - E_z^{n+1/2}(i,j+1)\right],$$

$$I_{Hx}^{n+1/2}\left(i,j+\frac{1}{2}\right) = I_{Hx}^{n-1/2}\left(i,j+\frac{1}{2}\right) + curl\_e,$$

$$H_x^{n+1}\left(i,j+\frac{1}{2}\right) = fj3\left(j+\frac{1}{2}\right)\cdot H_x^n\left(i,j+\frac{1}{2}\right) + fj2\left(j+\frac{1}{2}\right)\cdot\left[0.5\cdot curl\_e + fi1(i)\cdot I_{Hx}^{n+1/2}\left(i,j+\frac{1}{2}\right)\right].$$

Now, the full set of parameters associated with the PML is the following:

$$fi1(i), fj1(j) \quad \text{from 0 to 0.333} \tag{3.25a}$$

$$fi2(i), gi2(i), fj2(j), gj2(j) \quad \text{from 1 to 0.75} \tag{3.25b}$$

$$fi3(i), gi3(i), fj3(j), gj3(j) \quad \text{from 1 to 0.5} \tag{3.25c}$$

Notice that we could simply turn off the PML in the main part of the problem space by setting *fi*1 and *fj*1 to 0, and the other parameters to 1. Because they are only one-dimensional parameters, they add very little to the memory requirements. However, I_{Hx} and I_{Hy} are two-dimensional parameters. Although memory requirements are not a main issue while we are in two dimensions, when we get to three dimensions, we will think twice before introducing two new parameters that are defined throughout the problem space, but are needed only in a small fraction of the space.

The PML is implemented in the program fd2d_3_2.py. Figure 3.4 illustrates the effectiveness of an eight-point PML with the source offset five cells from center in both the x and y directions. Note that the outgoing contours remain concentric. Only when the wave gets within eight points of the edge, which is inside the PML, does distortion begin.

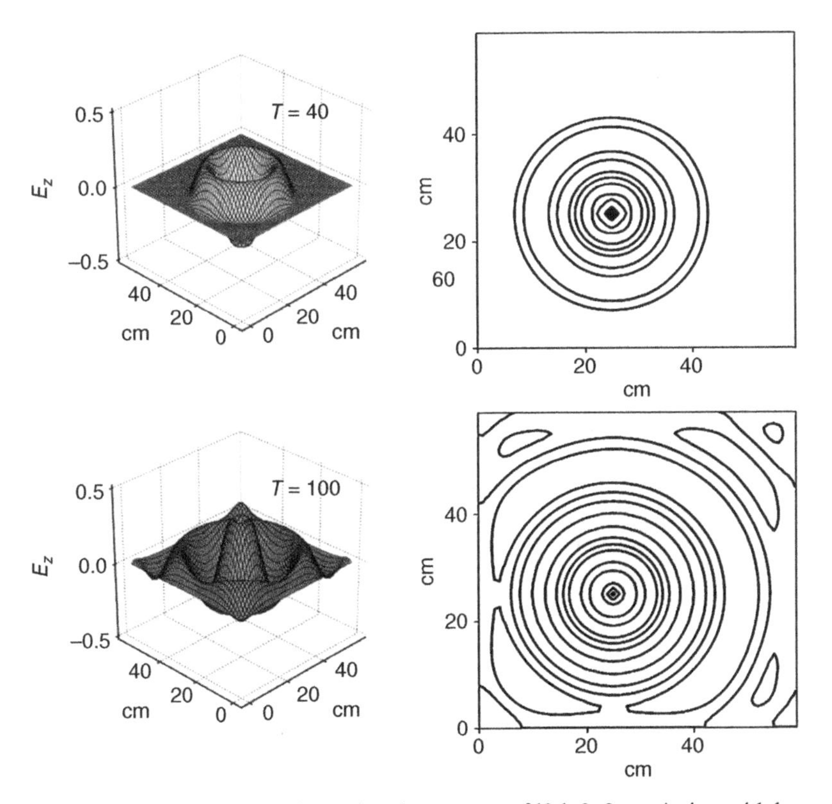

Figure 3.4 Results of a simulation using the program fd2d_3_2.py. A sinusoidal source is initiated at a point that is offset five cells from the center of the problem space in each direction. As the wave reaches the PML, which is eight cells on every side, it is absorbed. The effectiveness of the PML is apparent in the bottom figure because the contours would not be concentric circles if the outgoing wave ware partially reflected.

PROBLEM SET 3.2

1. The program fd2d_3_2.py is the same as fd2d_3_1.py, but with the two-dimensional PML added. Add the PML to your version of fd2d_3_1.py. Offset each point source by setting `ic = ie / 2 - 5` and `jc = je / 2 - 5`. Verify the results in Fig. 3.4.

3.3 TOTAL/SCATTERED FIELD FORMULATION

The simulation of plane waves is often of interest in computational electromagnetics. Many problems, such as the calculation of radar cross sections (2, 3), deal with plane waves. Furthermore, after a distance on the order of tens of wavelengths, the field from most antennas can be approximated as a plane wave.

In order to simulate a plane wave in a two-dimensional FDTD program, the problem space is divided into two regions, the *total field* and the *scattered field* (Fig. 3.5). The two primary reasons for doing this are (i) the propagating plane wave should not interact with the ABCs and (ii) the load on the ABCs should be minimized. These boundary conditions are not perfect, that is, a certain portion of the impinging wave is reflected back into the problem space. By subtracting the incident field, the amount of the radiating field hitting the boundary is minimized, thereby reducing the amount of error.

Figure 3.5 illustrates how this is accomplished. First, note that there is an auxiliary one-dimensional array called the *incident array*. Because it is a one-dimensional array, a plane wave can easily be generated: a source point is chosen and the incident E_z field is added at that point. Then a plane wave

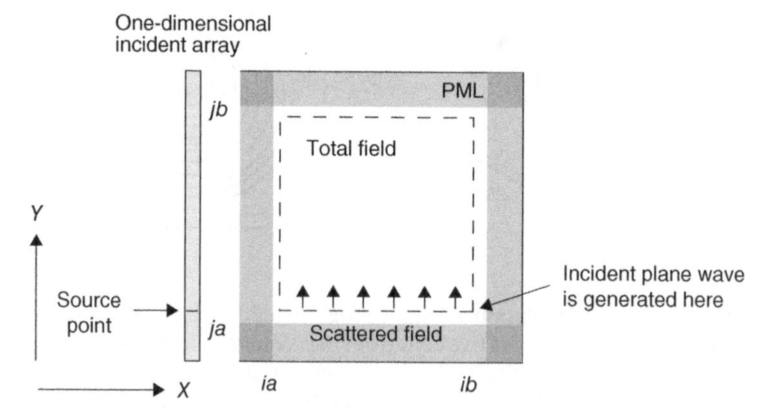

Figure 3.5 Total field/scattered field of the two-dimensional problem space.

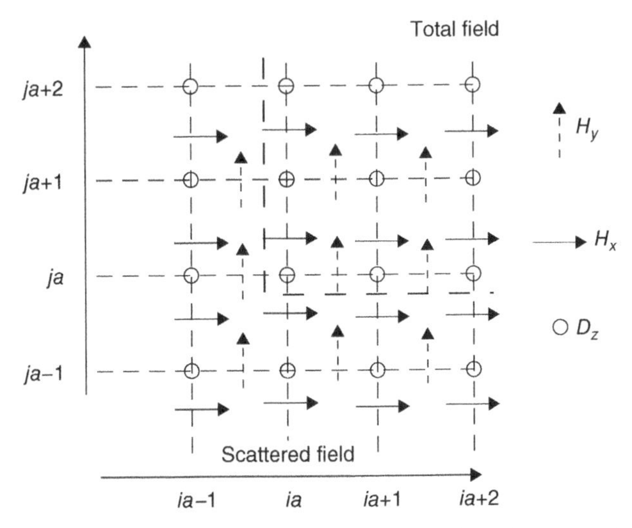

Figure 3.6 Every point is in either the total field or the scattered field.

propagates away in both directions. Since it is a one-dimensional array, the boundary conditions are perfect.

In the two-dimensional field, every point in the problem space is either in the total field or it is not; no point lies on the border (see Fig. 3.6). Therefore, if a point is within the total field but it uses points outside to calculate the spatial derivatives when updating its value, the point must be modified. The same is true of a point lying just outside that uses points inside the total field. The incident array contains the needed values to make these modifications.

Three places must be modified:

1. The D_z value at $j = j_a$ or $j = j_b$:

$$D_z(i, j_a) = D_z(i, j_a) + 0.5 \cdot H_{x\text{inc}}\left(j_a - \frac{1}{2}\right), \tag{3.26a}$$

$$D_z(i, j_b) = D_z(i, j_b) - 0.5 \cdot H_{x\text{inc}}\left(j_b + \frac{1}{2}\right). \tag{3.26b}$$

2. The H_x field just outside at $j = j_a$ or $j = j_b$:

$$H_x\left(i, j_a - \frac{1}{2}\right) = H_x\left(i, j_a - \frac{1}{2}\right) + 0.5 \cdot E_{z\text{inc}}(j_a), \tag{3.27a}$$

$$H_x\left(i, j_b + \frac{1}{2}\right) = H_x\left(i, j_b + \frac{1}{2}\right) - 0.5 \cdot E_{z\text{inc}}(j_b). \tag{3.27b}$$

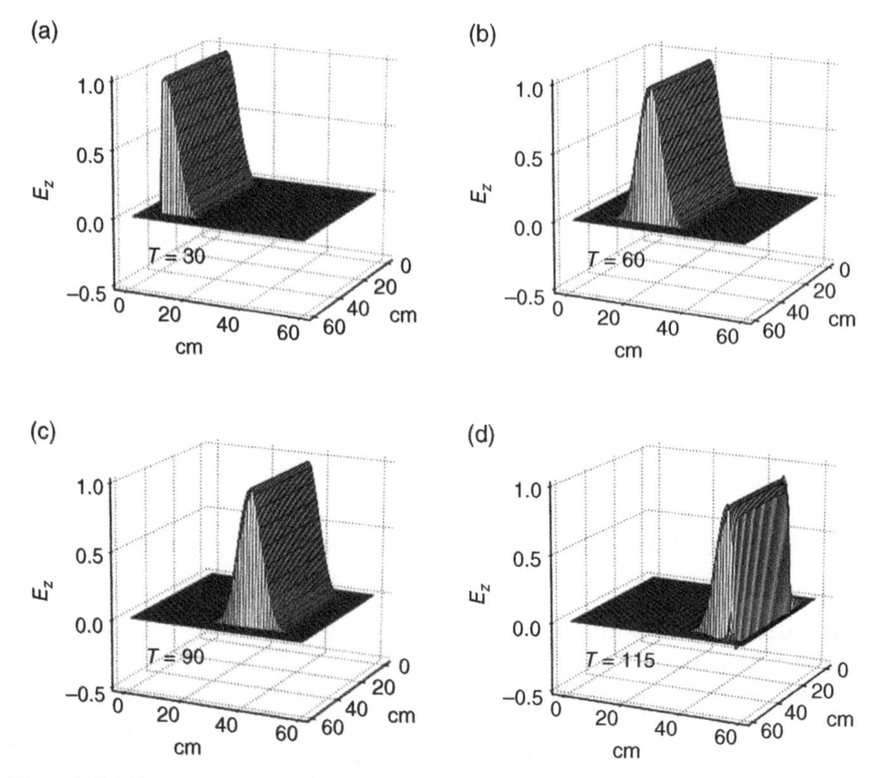

Figure 3.7 Simulation of a plane wave pulse propagating in free space. The incident pulse is generated at one end and subtracted from the other end. (a) T = 30, (b) T = 60, (c) T = 90, and (d) T = 115.

3. The H_y field just outside $i = i_a$ and $i = i_b$:

$$H_y\left(i_a - \frac{1}{2}, j\right) = H_y\left(i_a - \frac{1}{2}, j\right) - 0.5 \cdot E_{\text{zinc}}(j), \tag{3.28a}$$

$$H_y\left(i_b + \frac{1}{2}, j\right) = H_y\left(i_b + \frac{1}{2}, j\right) + 0.5 \cdot E_{\text{zinc}}(j). \tag{3.28b}$$

Figure 3.7 illustrates the propagation of a Gaussian pulse through the problem space. Notice how the pulse is generated at one end and completely subtracted from the other end.

3.3.1 A Plane Wave Impinging on a Dielectric Cylinder

Now we can simulate a plane wave. To simulate a plane wave interacting with an object, we must specify the object according to its electromagnetic properties: the dielectric constant and the conductivity. For instance, suppose we are simulating a plane wave striking a dielectric cylinder 20 cm in diameter, which has a dielectric constant specified by the parameter `epsr` and a conductivity specified by the parameter `sigma`. The cylinder is specified by the following Python code:

```
for j in range(ja, jb):
    for i in range(ia, ib):
        xdist = (ic - i)
        ydist = (jc - j)
        dist = sqrt(xdist ** 2 + ydist ** 2)
        if dist <= radius:
             gaz[i, j] = 1 / (epsr + (sigma * dt / epsz))
             gbz[i, j] = (sigma * dt / epsz)
```

Of course, this assumes that the problem space is initialized to free space. For every cell, the distance to the center of the problem space is calculated, and if it is less than the radius, the dielectric constant and conductivity are reset to `epsr` and `sigma`, respectively. A diagram of the problem space to simulate a plane wave interacting with a dielectric cylinder is shown in Fig. 3.8.

The weaknesses of this approach are obvious: Since we determined the properties by a simple "in-or-out" approach, we are left with the "staircasing" at the edge of the cylinder. It would be better if we had a way to make a smooth transition from one medium to another. One method is to divide each FDTD cell into subcells and then determine the average dielectric properties according to the number of subcells in one medium as well as in the other medium. The following code implements such a procedure:

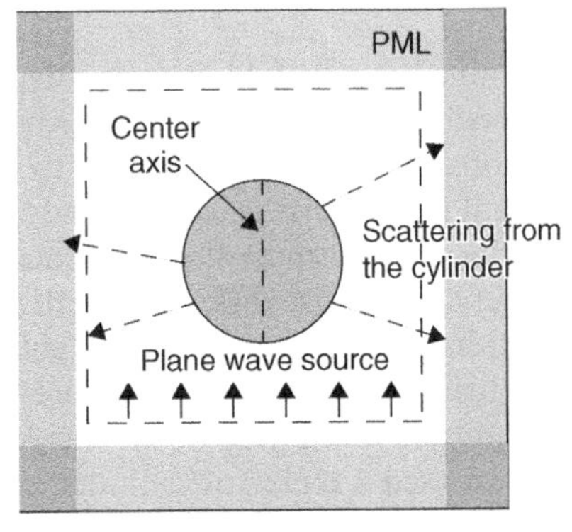

Figure 3.8 Simulation of a plane wave striking a dielectric cylinder. The fields scattered from the cylinder are the only fields to leave the total field and strike the PML.

```
for j in range(ja, jb):
    for i in range(ia, ib):
        eps = 1
        cond = 0
        for jj in range(-1, 2):
            for ii in range(-1, 2):
                xdist = (ic - i) + 1 / 3 * ii
                ydist = (jc - j) + 1 / 3 * jj
                dist = sqrt(xdist ** 2 + ydist ** 2)
                if dist <= radius:
                    eps = eps + (1 / 9) * (epsr - 1)
                    cond = cond + (1 / 9) * (sigma)
        gaz[i, j] = 1.0 / (eps + cond * dt / epsz)
        gbz[i, j] = cond * dt / epsz
```

Each cell is initialized to the values `eps = 1`, `cond = 0`. Notice that each of the inner loops, which are iterated by the parameters `ii` and `jj`, move the distance one-third of a cell length. Thus, the cell has been divided into nine subcells. Also, notice that when a subcell is determined to be within the radius, its contribution (`epsr`) is added to the total epsilon, while subtracting out the initialized epsilon. The above code may be modified to create an arbitrary number of layers in the cylinder by following the same pattern.

The simulation of a plane wave pulse hitting a dielectric cylinder with $\varepsilon_r = 30$ and $\sigma = 0.3$ is shown in Fig. 3.9. After 25 time steps, the plane wave has started from the side; after 50 time steps, the pulse is interacting with the cylinder. Some of the pulse passes through the cylinder, and some of it goes around it. After 100 steps, the main part of the propagating pulse is being subtracted from the end of the total field.

3.3.2 Fourier Analysis

Suppose we want to determine how the EM energy is deposited within the cylinder for a plane wave propagating at a specific frequency. Recall that we did something similar in Section 2.3 where we were able to determine the attenuation of the E field at several frequencies by using a pulse for a source and calculating the discrete Fourier transform at the frequencies of interest. This is exactly what we will do here. The only difference is that because it is a two-dimensional space, we will have a larger number of points. Furthermore, we can use the Fourier transform of the pulse in the one-dimensional incident buffer to calculate the amplitude and phase of the incident pulse.

There is a reason why we used a dielectric cylinder as the object: It has an analytical solution. The fields resulting from a plane wave at a single frequency interacting with a dielectric cylinder can be calculated through a Bessel function expansion (8). This gives us an opportunity to check the accuracy of our simulations. Figure 3.10 is a comparison of the FDTD calculations with the analytic solutions from Bessel function expansion along the center axis of the cylinder in

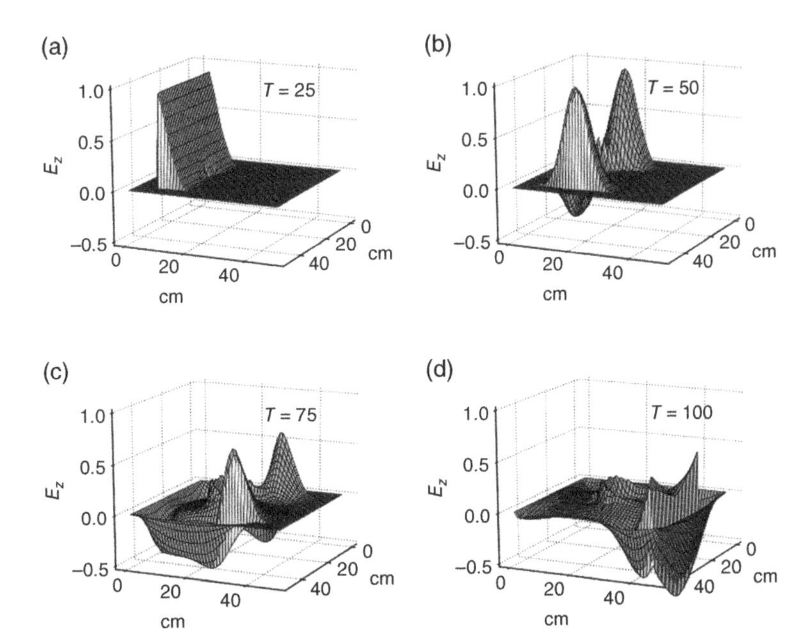

Figure 3.9 Simulation of a plane wave impinging on a dielectric cylinder. The cylinder is 20 cm in diameter and has a relative dielectric constant of 30 and a conductivity of 0.3 S/m. (a) $T = 25$, (b) $T = 50$, (c) $T = 75$, and (d) $T = 100$.

the propagation direction (see Fig. 3.8) at 50, 300, and 700 MHz. This was calculated with the program that averaged the values across the boundaries, as described earlier. The accuracy is quite good. Remember, we are able to calculate all three frequencies with a single computer run by using the impulse response method.

PROBLEMS SET 3.3

1. The program fd2d_3_3.py implements the two-dimensional TM FDTD algorithm with an incident plane wave. Note that Eq. (3.26) is implemented to give the correct E_z field at the total/scattered field boundary and Eq. (3.27) is implemented to give the correct H_x field. Why is there no modification of the H_y field?

2. Get the program fd2d_3_3.py running. You should be able to observe the pulse that is generated at `j = ja`, propagates through the problem space, and is subtracted out at `j = jb`.

3. The program fd2d_3_4.py differs from fd2d_3_3.py because it simulates a plane wave hitting a dielectric cylinder. It also calculates the frequency response at three frequencies within the cylinder. Get this running and verify the results in Fig. 3.10.

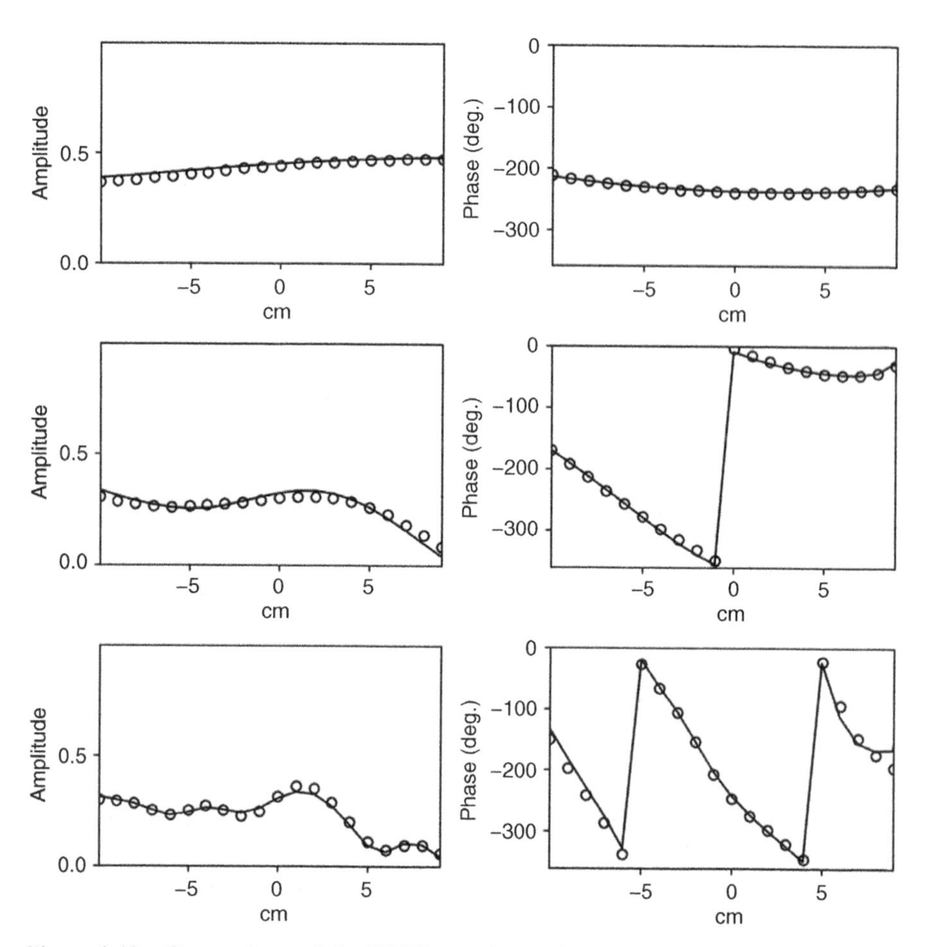

Figure 3.10 Comparison of the FDTD results (solid lines) with the Bessel function expansion results (circles) along the propagation center axis of a cylinder at three frequencies. The cylinder is 20 cm in diameter and has a relative dielectric constant of 30 and a conductivity of 0.3 S/m.

4. The cylinder generated in fd2d_3_4.py generates the cylinder by the "in-or-out" method. Change this to the averaged values and add the ability to generate layered cylinders, as described in Section 3.3.1.

REFERENCES

1. K. S. Yee, Numerical solution of initial boundary value problems involving Maxwell's equations, in isotropic media, *IEEE Trans. Antennas Propag.*, vol. AP-17, 1996, pp. 585–589.
2. A. Taflove, *Computational Electrodynamics: The Finite-Difference Time-Domain Method.* Boston, MA: Artech House, 1995.

3. K. S. Kunz, and R. J. Luebbers, *The Finite Difference Time Domain Method for Electromagnetics*. Boca Raton, FL: CRC Press, 1993.
4. J. P. Berenger, A perfectly matched layer for the absorption of electromagnetic waves, *J. Comput. Phys.*, vol. 114, 1994, pp. 185–200.
5. D. M. Sullivan, A simplified PML for use with the FDTD method, *IEEE Microw. Guid. Wave Lett.*, vol. 6, Feb. 1996, pp. 97–99.
6. Z. S. Sacks, D. M. Kingsland, T. Lee, and J. F. Lee, A perfectly matched anisotropic absorber for use as an absorbing boundary condition, *IEEE Trans. Antennas Propag.*, vol. 43, Dec. 1995, pp. 1460–1463.
7. D. M. Sullivan, An unsplit step 3D PML for use with the FDTD Method, *IEEE Microw. Guid. Wave Lett.*, vol. 7, Feb. 1997, pp. 184–186.
8. R. Harrington, *Time-Harmonic Electromagnetic Fields*, New York: McGraw-Hill, 1961.

```
""" fd2d_3_1.py: 2D FDTD

TM program
"""

import numpy as np
from math import exp
from matplotlib import pyplot as plt
import mpl_toolkits.mplot3d.axes3d

ie = 60
je = 60
ic = int(ie / 2)
jc = int(je / 2)

ez = np.zeros((ie, je))
dz = np.zeros((ie, je))
hx = np.zeros((ie, je))
hy = np.zeros((ie, je))

ddx = 0.01  # Cell size
dt = ddx / 6e8  # Time step size

# Create Dielectric Profile
epsz = 8.854e-12

# Pulse Parameters
t0 = 20
spread = 6

gaz = np.ones((ie, je))

nsteps = 50
```

```
# Dictionary to keep track of desired points for plotting
plotting_points = [
    {'label': 'a', 'num_steps': 20, 'data_to_plot': None},
    {'label': 'b', 'num_steps': 30, 'data_to_plot': None},
    {'label': 'c', 'num_steps': 40, 'data_to_plot': None},
    {'label': 'd', 'num_steps': 50, 'data_to_plot': None},
]

# Main FDTD Loop
for time_step in range(1, nsteps + 1):

    # Calculate Dz
    for j in range(1, je):
        for i in range(1, ie):
            dz[i, j] = dz[i, j] + 0.5 * (hy[i, j] - hy[i - 1, j] -
                                         hx[i, j] + hx[i, j - 1])

    # Put a Gaussian pulse in the middle
    pulse = exp(-0.5 * ((t0 - time_step) / spread) ** 2)
    dz[ic, jc] = pulse

    # Calculate the Ez field from Dz
    for j in range(1, je):
        for i in range(1, ie):
            ez[i, j] = gaz[i, j] * dz[i, j]

    # Calculate the Hx field
    for j in range(je - 1):
        for i in range(ie - 1):
            hx[i, j] = hx[i, j] + 0.5 * (ez[i, j] - ez[i, j + 1])

    # Calculate the Hy field
    for j in range(je - 1):
        for i in range(ie - 1):
             hy[i, j] = hy[i, j] + 0.5 * (ez[i + 1, j] - ez[i, j])

    # Save data at certain points for later plotting
    for plotting_point in plotting_points:
        if time_step == plotting_point['num_steps']:
           plotting_point['data_to_plot'] = np.copy(ez)

# Plot Fig. 3.2
plt.rcParams['font.size'] = 12
plt.rcParams['grid.color'] = 'gray'
plt.rcParams['grid.linestyle'] = 'dotted'
fig = plt.figure(figsize=(8, 7))

X, Y = np.meshgrid(range(je), range(ie))
```

```
def plot_e_field(ax, data, timestep, label):
    """3d Plot of E field at a single time step"""
    ax.set_zlim(0, 1)
    ax.view_init(elev=20., azim=45)
    ax.plot_surface(X, Y, data[:, :], rstride=1, cstride=1,
                    color='white',
                    edgecolor='black', linewidth=.25)
    ax.zaxis.set_rotate_label(False)
    ax.set_zlabel(r' $E_{Z}$', rotation=90, labelpad=10,
                  fontsize=14)
    ax.set_zticks([0, 0.5, 1])
    ax.set_xlabel('cm')
    ax.set_ylabel('cm')
    ax.set_xticks(np.arange(0, 61, step=20))
    ax.set_yticks(np.arange(0, 61, step=20))
    ax.text2D(0.6, 0.7, "T = {}".format(timestep), transform=ax.
          transAxes)
    ax.xaxis.pane.fill = ax.yaxis.pane.fill = ax.zaxis.pane.fill
     = False
    plt.gca().patch.set_facecolor('white')
    ax.text2D(-0.2, 0.8, "({})".format(label), transform=ax.
          transAxes)
    ax.dist = 11

# Plot the E field at each of the four time steps saved earlier
for subplot_num, plotting_point in enumerate(plotting_points):
    ax = fig.add_subplot(2, 2, subplot_num + 1, projection='3d')
    plot_e_field(ax, plotting_point['data_to_plot'],
                 plotting_point['num_steps'],
                 plotting_point['label'])

plt.subplots_adjust(bottom=0.05, left=0.10, hspace=0.05)
plt.show()


""" fd2d_3_2.py: 2D FDTD

TM program with the PML added
"""

import numpy as np
from math import sin, pi
from matplotlib import pyplot as plt
import mpl_toolkits.mplot3d.axes3d

ie = 60
je = 60
ic = int(ie / 2 - 5)
jc = int(je / 2 - 5)
```

```
ez = np.zeros((ie, je))
dz = np.zeros((ie, je))
hx = np.zeros((ie, je))
hy = np.zeros((ie, je))
ihx = np.zeros((ie, je))
ihy = np.zeros((ie, je))

ddx = 0.01  # Cell size
dt = ddx / 6e8  # Time step size

# Create Dielectric Profile
epsz = 8.854e-12

# Pulse Parameters
t0 = 40
spread = 12

gaz = np.ones((ie, je))

# Calculate the PML parameters
gi2 = np.ones(ie)
gi3 = np.ones(ie)
fi1 = np.zeros(ie)
fi2 = np.ones(ie)
fi3 = np.ones(ie)

gj2 = np.ones(ie)
gj3 = np.ones(ie)
fj1 = np.zeros(ie)
fj2 = np.ones(ie)
fj3 = np.ones(ie)

# Create the PML as described in Section 3.2
npml = 8
for n in range(npml):
    xnum = npml - n
    xd = npml
    xxn = xnum / xd
    xn = 0.33 * xxn ** 3

    gi2[n] = 1 / (1 + xn)
    gi2[ie - 1 - n] = 1 / (1 + xn)
    gi3[n] = (1 - xn) / (1 + xn)
    gi3[ie - 1 - n] = (1 - xn) / (1 + xn)

    gj2[n] = 1 / (1 + xn)
    gj2[je - 1 - n] = 1 / (1 + xn)
    gj3[n] = (1 - xn) / (1 + xn)
    gj3[je - 1 - n] = (1 - xn) / (1 + xn)
```

```
    xxn = (xnum - 0.5) / xd
    xn = 0.33 * xxn ** 3

    fi1[n] = xn
    fi1[ie - 2 - n] = xn
    fi2[n] = 1 / (1 + xn)
    fi2[ie - 2 - n] = 1 / (1 + xn)
    fi3[n] = (1 - xn) / (1 + xn)
    fi3[ie - 2 - n] = (1 - xn) / (1 + xn)

    fj1[n] = xn
    fj1[je - 2 - n] = xn
    fj2[n] = 1 / (1 + xn)
    fj2[je - 2 - n] = 1 / (1 + xn)
    fj3[n] = (1 - xn) / (1 + xn)
    fj3[je - 2 - n] = (1 - xn) / (1 + xn)

nsteps = 100

# Dictionary to keep track of desired points for plotting
plotting_points = [
    {'num_steps': 40, 'data_to_plot': None},
    {'num_steps': nsteps, 'data_to_plot': None},
]

# Main FDTD Loop
for time_step in range(1, nsteps + 1):

    # Calculate Dz
    for j in range(1, je):
        for i in range(1, ie):
            dz[i, j] = gi3[i] * gj3[j] * dz[i, j] + \
                       gi2[i] * gj2[j] * 0.5 * \
                       (hy[i, j] - hy[i - 1, j] -
                        hx[i, j] + hx[i, j - 1])

    # Put a Gaussian pulse in the middle
    pulse = sin(2 * pi * 1500 * 1e6 * dt * time_step)
    dz[ic, jc] = pulse

    ez = gaz * dz  # Calculate the Ez field from Dz

    # Calculate the Hx field
    for j in range(je - 1):
        for i in range(ie - 1):
            curl_e = ez[i, j] - ez[i, j + 1]
            ihx[i, j] = ihx[i, j] + curl_e
            hx[i, j] = fj3[j] * hx[i, j] + fj2[j] * \
                       (0.5 * curl_e + fi1[i] * ihx[i, j])
```

```
    # Calculate the Hy field
    for j in range(0, je - 1):
        for i in range(0, ie - 1):
            curl_e = ez[i, j] - ez[i + 1, j]
            ihy[i, j] = ihy[i, j] + curl_e
            hy[i, j] = fi3[i] * hy[i, j] - fi2[i] * \
                       (0.5 * curl_e + fj1[j] * ihy[i, j])

    # Save data at certain points for later plotting
    for plotting_point in plotting_points:
        if time_step == plotting_point['num_steps']:
            plotting_point['data_to_plot'] = np.copy(ez)

# Plot Fig. 3.4
plt.rcParams['font.size'] = 12
plt.rcParams['grid.color'] = 'gray'
plt.rcParams['grid.linestyle'] = 'dotted'
fig = plt.figure(figsize=(8, 8))

X, Y = np.meshgrid(range(je), range(ie))


def plot_e_field(ax, data, timestep):
    """3d Plot of E field at a single time step"""
    ax.set_zlim(-0.5, 0.5)
    ax.view_init(elev=30., azim=-135)
    ax.plot_surface(X, Y, data, rstride=1, cstride=1,
                    color='white',
                    edgecolor='black', linewidth=.25)
    ax.zaxis.set_rotate_label(False)
    ax.set_zlabel(r' $E_{Z}$', rotation=90, labelpad=10,
                  fontsize=14)
    ax.set_xlabel('cm')
    ax.set_ylabel('cm')
    ax.set_xticks(np.arange(0, 60, step=20))
    ax.set_yticks(np.arange(0, 60, step=20))
    ax.set_zticks([-0.5, 0, 0.5])
    ax.text2D(0.6, 0.7, "T = {}".format(timestep),
              transform=ax.transAxes)
    ax.xaxis.pane.fill = ax.yaxis.pane.fill = \
        ax.zaxis.pane.fill = False
    plt.gca().patch.set_facecolor('white')
    ax.dist = 11


def plot_e_field_contour(ax, data):
    """Contour Plot of E field at a single time step"""
    CP = plt.contour(X, Y, data, colors='black',
                     linestyles='solid')
```

```
    CP.collections[4].remove()
    # above removes extraneous outer contour display
    ax.set_xticks(np.arange(0, 60, step=20))
    ax.set_yticks(np.arange(0, 60, step=20))
    plt.xlabel('cm')
    plt.ylabel('cm')


# Plot the E field at each of the time steps saved earlier
for subplot_num, plotting_point in enumerate(plotting_points):
    ax = fig.add_subplot(2, 2, subplot_num * 2 + 1,
                         projection='3d')
    plot_e_field(ax, plotting_point['data_to_plot'],
                 plotting_point['num_steps'])
    ax = fig.add_subplot(2, 2, subplot_num * 2 + 2)
    plot_e_field_contour(ax, plotting_point['data_to_plot'])

plt.tight_layout()
plt.subplots_adjust(left=0.05)
plt.show()

""" fd2d_3_3.py: 2D FDTD

TM program with plane wave source
"""

import numpy as np
from math import exp
from matplotlib import pyplot as plt
from mpl_toolkits.mplot3d.axes3d import Axes3D, get_test_data

ie = 60
je = 60
ic = int(ie / 2)
jc = int(je / 2)
ia = 7
ib = ie - ia - 1
ja = 7
jb = je - ja - 1

ez = np.zeros((ie, je))
dz = np.zeros((ie, je))
hx = np.zeros((ie, je))
hy = np.zeros((ie, je))
ihx = np.zeros((ie, je))
ihy = np.zeros((ie, je))

ez_inc = np.zeros(je)
hx_inc = np.zeros(je)
```

```
ddx = 0.01  # Cell size
dt = ddx / 6e8  # Time step size

# Absorbing Boundary Conditions
boundary_low = [0, 0]
boundary_high = [0, 0]

# Calculate the PML parameters
gi2 = np.ones(ie)
gi3 = np.ones(ie)
fi1 = np.zeros(ie)
fi2 = np.ones(ie)
fi3 = np.ones(ie)

gj2 = np.ones(ie)
gj3 = np.ones(ie)
fj1 = np.zeros(ie)
fj2 = np.ones(ie)
fj3 = np.ones(ie)

# Create the PML as described in Section 3.2
npml = 8
for n in range(npml):
    xnum = npml - n
    xd = npml
    xxn = xnum / xd
    xn = 0.33 * xxn ** 3

    gi2[n] = 1 / (1 + xn)
    gi2[ie - 1 - n] = 1 / (1 + xn)
    gi3[n] = (1 - xn) / (1 + xn)
    gi3[ie - 1 - n] = (1 - xn) / (1 + xn)

    gj2[n] = 1 / (1 + xn)
    gj2[je - 1 - n] = 1 / (1 + xn)
    gj3[n] = (1 - xn) / (1 + xn)
    gj3[je - 1 - n] = (1 - xn) / (1 + xn)

    xxn = (xnum - 0.5) / xd
    xn = 0.33 * xxn ** 3

    fi1[n] = xn
    fi1[ie - 2 - n] = xn
    fi2[n] = 1 / (1 + xn)
    fi2[ie - 2 - n] = 1 / (1 + xn)
    fi3[n] = (1 - xn) / (1 + xn)
    fi3[ie - 2 - n] = (1 - xn) / (1 + xn)

    fj1[n] = xn
    fj1[je - 2 - n] = xn
    fj2[n] = 1 / (1 + xn)
```

```
    fj2[je - 2 - n] = 1 / (1 + xn)
    fj3[n] = (1 - xn) / (1 + xn)
    fj3[je - 2 - n] = (1 - xn) / (1 + xn)

# Create Dielectric Profile
epsz = 8.854e-12

# Pulse Parameters
t0 = 20
spread = 8

gaz = np.ones((ie, je))

nsteps = 115

# Dictionary to keep track of desired points for plotting
plotting_points = [
    {'label': 'a', 'num_steps': 30, 'data_to_plot': None},
    {'label': 'b', 'num_steps': 60, 'data_to_plot': None},
    {'label': 'c', 'num_steps': 90, 'data_to_plot': None},
    {'label': 'd', 'num_steps': 115, 'data_to_plot': None},
]

# Main FDTD Loop
for time_step in range(1, nsteps + 1):

    for j in range(1, je):
        ez_inc[j] = ez_inc[j] + 0.5 * (hx_inc[j - 1] - hx_inc[j])

    # Absorbing Boundary Conditions
    ez_inc[0] = boundary_low.pop(0)
    boundary_low.append(ez_inc[1])

    ez_inc[je - 1] = boundary_high.pop(0)
    boundary_high.append(ez_inc[je - 2])

    # Calculate the Dz field
    for j in range(1, je):
        for i in range(1, ie):
            dz[i, j] = gi3[i] * gj3[j] * dz[i, j] + \
                       gi2[i] * gj2[j] * 0.5 * \
                       (hy[i, j] - hy[i - 1, j] -
                        hx[i, j] + hx[i, j - 1])

    # Source
    pulse = exp(-0.5 * ((t0 - time_step) / spread) ** 2)
    ez_inc[3] = pulse
```

```
    # Incident Dz values
    for i in range(ia, ib):
        dz[i, ja] = dz[i, ja] + 0.5 * hx_inc[ja - 1]
        dz[i, jb] = dz[i, jb] - 0.5 * hx_inc[jb - 1]

    ez = gaz * dz  # Calculate the Ez field from Dz

    for j in range(0, je - 1):
        hx_inc[j] = hx_inc[j] + 0.5 * (ez_inc[j] - ez_inc[j + 1])

    for j in range(0, je - 1):
        for i in range(0, ie - 1):
            curl_e = ez[i, j] - ez[i, j + 1]
            ihx[i, j] = ihx[i, j] + curl_e
            hx[i, j] = fj3[j] * hx[i, j] + fj2[j] * \
                       (0.5 * curl_e + fi1[i] * ihx[i, j])

    # Incident Hx values
    for i in range(ia, ib):
        hx[i, ja - 1] = hx[i, ja - 1] + 0.5 * ez_inc[ja]
        hx[i, jb] = hx[i, jb] - 0.5 * ez_inc[jb]

    # Calculate the Hy field
    for j in range(0, je - 1):
        for i in range(0, ie - 1):
            curl_e = ez[i, j] - ez[i + 1, j]
            ihy[i, j] = ihy[i, j] + curl_e
            hy[i, j] = fi3[i] * hy[i, j] - fi2[i] * \
                       (0.5 * curl_e + fj1[j] * ihy[i, j])

    # Incident Hy values
    for j in range(ja, jb):
        hy[ia - 1, j] = hy[ia - 1, j] - 0.5 * ez_inc[j]
        hy[ib - 1, j] = hy[ib - 1, j] + 0.5 * ez_inc[j]

    # Save data at certain points for later plotting
    for plotting_point in plotting_points:
        if time_step == plotting_point['num_steps']:
           plotting_point['data_to_plot'] = np.copy(ez)

# Plot Fig. 3.7
plt.rcParams['font.size'] = 12
plt.rcParams['grid.color'] = 'gray'
plt.rcParams['grid.linestyle'] = 'dotted'
fig = plt.figure(figsize=(8, 8))

X, Y = np.meshgrid(range(je), range(ie))
```

```
def plot_e_field(ax, data, timestep, label):
    """3d Plot of E field at a single timestep"""
    ax.set_zlim(-0.5, 1)
    ax.view_init(elev=15., azim=25)
    ax.plot_surface(Y, X, data, rstride=1, cstride=1,
                    color='white',
                    edgecolor='black', linewidth=.25)
    ax.zaxis.set_rotate_label(False)
    ax.set_zlabel(r' $E_{Z}$', rotation=90,
                  labelpad=10, fontsize=14)
    ax.set_zticks([-0.5, 0, 0.5, 1])
    ax.set_xlabel('cm')
    ax.set_ylabel('cm')
    ax.set_xticks(np.arange(0, 61, step=20))
    ax.set_yticks(np.arange(0, 61, step=20))
    ax.text2D(0.25, 0.3, "T = {}".format(timestep),
              transform=ax.transAxes)
    ax.xaxis.pane.fill = ax.yaxis.pane.fill = \
        ax.zaxis.pane.fill = False
    plt.gca().patch.set_facecolor('white')
    ax.text2D(-0.05, 0.8, "({})".format(label),
              transform=ax.transAxes)
    ax.dist = 11


# Plot the E field at each of the four time steps saved earlier
for subplot_num, plotting_point in enumerate(plotting_points):
    ax = fig.add_subplot(2, 2, subplot_num + 1, projection='3d')
    plot_e_field(ax, plotting_point['data_to_plot'],
                 plotting_point['num_steps'], plotting_point
                   ['label'])

fig.tight_layout()
plt.show()


""" fd2d_3_4.py: 2D FDTD

TM simulation of a plane wave source impinging on a dielectric
  cylinder
Analysis using Fourier transforms
"""

import numpy as np
from math import sin, exp, sqrt, atan2, cos, pi
from matplotlib import pyplot as plt
from mpl_toolkits.mplot3d.axes3d import Axes3D, get_test_data
```

```
ie = 50
je = 50
ic = int(ie / 2 - 1)
jc = int(je / 2 - 1)
ia = 7
ib = ie - ia - 1
ja = 7
jb = je - ja - 1

ez = np.zeros((ie, je))
dz = np.zeros((ie, je))
hx = np.zeros((ie, je))
hy = np.zeros((ie, je))
iz = np.zeros((ie, je))
ihx = np.zeros((ie, je))
ihy = np.zeros((ie, je))

ez_inc = np.zeros(je)
hx_inc = np.zeros(je)

ddx = 0.01  # Cell size
dt = ddx / 6e8  # Time step size

number_of_frequencies = 3
freq = np.array((50e6, 300e6, 700e6))
arg = 2 * np.pi * freq * dt
real_in = np.zeros(number_of_frequencies)
imag_in = np.zeros(number_of_frequencies)
real_pt = np.zeros((number_of_frequencies, ie, je))
imag_pt = np.zeros((number_of_frequencies, ie, je))
amp = np.zeros((number_of_frequencies, je))
phase = np.zeros((number_of_frequencies, je))
gaz = np.ones((ie, je))
gbz = np.zeros((ie, je))

# Specify the dielectric cylinder
epsr = 30
sigma = 0.3
radius = 10

# Create Dielectric Profile
epsz = 8.854e-12

for j in range(ja, jb):
    for i in range(ia, ib):
        xdist = (ic - i)
        ydist = (jc - j)
        dist = sqrt(xdist ** 2 + ydist ** 2)
        if dist <= radius:
```

```
            gaz[i, j] = 1 / (epsr + (sigma * dt / epsz))
            gbz[i, j] = (sigma * dt / epsz)
boundary_low = [0, 0]
boundary_high = [0, 0]

# Calculate the PML parameters
gi2 = np.ones(ie)
gi3 = np.ones(ie)
fi1 = np.zeros(ie)
fi2 = np.ones(ie)
fi3 = np.ones(ie)

gj2 = np.ones(je)
gj3 = np.ones(je)
fj1 = np.zeros(je)
fj2 = np.ones(je)
fj3 = np.ones(je)

# Create the PML as described in Section 3.2
npml = 8
for n in range(npml):
    xnum = npml - n
    xd = npml
    xxn = xnum / xd
    xn = 0.33 * xxn ** 3

    gi2[n] = 1 / (1 + xn)
    gi2[ie - 1 - n] = 1 / (1 + xn)
    gi3[n] = (1 - xn) / (1 + xn)
    gi3[ie - 1 - n] = (1 - xn) / (1 + xn)

    gj2[n] = 1 / (1 + xn)
    gj2[je - 1 - n] = 1 / (1 + xn)
    gj3[n] = (1 - xn) / (1 + xn)
    gj3[je - 1 - n] = (1 - xn) / (1 + xn)

    xxn = (xnum - 0.5) / xd
    xn = 0.33 * xxn ** 3

    fi1[n] = xn
    fi1[ie - 2 - n] = xn
    fi2[n] = 1 / (1 + xn)
    fi2[ie - 2 - n] = 1 / (1 + xn)
    fi3[n] = (1 - xn) / (1 + xn)
    fi3[ie - 2 - n] = (1 - xn) / (1 + xn)

    fj1[n] = xn
    fj1[je - 2 - n] = xn
    fj2[n] = 1 / (1 + xn)
```

```
    fj2[je - 2 - n] = 1 / (1 + xn)
    fj3[n] = (1 - xn) / (1 + xn)
    fj3[je - 2 - n] = (1 - xn) / (1 + xn)

# Pulse Parameters
t0 = 20
spread = 8

nsteps = 500

# Dictionary to keep track of desired points for plotting
plotting_points = [
    {'label': 'a', 'num_steps': 25, 'data_to_plot': None},
    {'label': 'b', 'num_steps': 50, 'data_to_plot': None},
    {'label': 'c', 'num_steps': 75, 'data_to_plot': None},
    {'label': 'd', 'num_steps': 100, 'data_to_plot': None},
]

# Main FDTD Loop
for time_step in range(1, nsteps + 1):

    # Incident Ez values
    for j in range(1, je):
        ez_inc[j] = ez_inc[j] + 0.5 * (hx_inc[j - 1] - hx_inc[j])

    if time_step < 3 * t0:
        for m in range(number_of_frequencies):
            real_in[m] = real_in[m] + cos(arg[m] * time_step) \
                         * ez_inc[ja - 1]
            imag_in[m] = imag_in[m] - sin(arg[m] * time_step) \
                         * ez_inc[ja - 1]

    # Absorbing Boundary Conditions
    ez_inc[0] = boundary_low.pop(0)
    boundary_low.append(ez_inc[1])

    ez_inc[je - 1] = boundary_high.pop(0)
    boundary_high.append(ez_inc[je - 2])

    # Calculate the Dz field
    for j in range(1, je):
        for i in range(1, ie):
            dz[i, j] = gi3[i] * gj3[j] * dz[i, j] + \
                       gi2[i] * gj2[j] * 0.5 * \
                       (hy[i, j] - hy[i - 1, j] -
                        hx[i, j] + hx[i, j - 1])
```

```
# Source
pulse = exp(-0.5 * ((t0 - time_step) / spread) ** 2)
ez_inc[3] = pulse

# Incident Dz values
for i in range(ia, ib + 1):
    dz[i, ja] = dz[i, ja] + 0.5 * hx_inc[ja - 1]
    dz[i, jb] = dz[i, jb] - 0.5 * hx_inc[jb]

# Calculate the Ez field
for j in range(0, je):
    for i in range(0, ie):
        ez[i, j] = gaz[i, j] * (dz[i, j] - iz[i, j])
        iz[i, j] = iz[i, j] + gbz[i, j] * ez[i, j]

# Calculate the Fourier transform of Ex
for j in range(0, je):
    for i in range(0, ie):
        for m in range(0, number_of_frequencies):
            real_pt[m, i, j] = real_pt[m, i, j] + \
                               cos(arg[m] * time_step) * ez[i, j]
            imag_pt[m, i, j] = imag_pt[m, i, j] - \
                               sin(arg[m] * time_step) * ez[i, j]

# Calculate the Incident Hx
for j in range(0, je - 1):
    hx_inc[j] = hx_inc[j] + 0.5 * (ez_inc[j] - ez_inc[j + 1])

# Calculate the Hx field
for j in range(0, je - 1):
    for i in range(0, ie - 1):
        curl_e = ez[i, j] - ez[i, j + 1]
        ihx[i, j] = ihx[i, j] + curl_e
        hx[i, j] = fj3[j] * hx[i, j] + fj2[j] * \
                   (0.5 * curl_e + fi1[i] * ihx[i, j])

# Incident Hx values
for i in range(ia, ib + 1):
    hx[i, ja - 1] = hx[i, ja - 1] + 0.5 * ez_inc[ja]
    hx[i, jb] = hx[i, jb] - 0.5 * ez_inc[jb]

# Calculate the Hy field
for j in range(0, je):
    for i in range(0, ie - 1):
        curl_e = ez[i, j] - ez[i + 1, j]
        ihy[i, j] = ihy[i, j] + curl_e
        hy[i, j] = fi3[i] * hy[i, j] - fi2[i] * \
                   (0.5 * curl_e + fj1[j] * ihy[i, j])
```

```
    # Incident Hy values
    for j in range(ja, jb + 1):
        hy[ia - 1, j] = hy[ia - 1, j] - 0.5 * ez_inc[j]
        hy[ib, j] = hy[ib, j] + 0.5 * ez_inc[j]

    # Save data at certain points for later plotting
    for plotting_point in plotting_points:
        if time_step == plotting_point['num_steps']:
           plotting_point['data_to_plot'] = np.copy(ez)

# Calculate the Fourier amplitude and phase of the incident pulse
amp_in = np.sqrt(real_in ** 2 + imag_in ** 2)
phase_in = np.arctan2(imag_in, real_in)

# This accounts for the offset between this program and the
  Bessel outputs
phase_offset = (0, 0.35 * pi, 0.9 * pi)

# Calculate the Fourier amplitude and phase of the total field
for m in range(number_of_frequencies):
    for j in range(ja, jb + 1):
        if gaz[ic, j] < 1:
           amp[m, j] = 1 / (amp_in[m]) * sqrt(real_pt[m, ic, j] ** 2 +
                                        imag_pt[m, ic, j] ** 2)
           phase[m, j] = atan2(imag_pt[m, ic, j], real_pt[m, ic, j]) \
                       - pi - phase_in[m] + phase_offset[m]
           if phase[m, j] < -2 * pi:
              phase[m, j] += 2 * pi
           if phase[m, j] > 0:
              phase[m, j] -= 2 * pi

# Plot Fig. 3.9
plt.rcParams['font.size'] = 12
plt.rcParams['grid.color'] = 'gray'
plt.rcParams['grid.linestyle'] = 'dotted'
fig = plt.figure(figsize=(8, 7))

X, Y = np.meshgrid(range(je), range(ie))


def plot_e_field(ax, data, timestep, label):
    """3d Plot of E field at a single timestep"""
    ax.set_zlim(-0.5, 1)
    ax.view_init(elev=15., azim=25)
    ax.plot_surface(Y, X, data, rstride=1, cstride=1,
                    color='white',
                    edgecolor='black', linewidth=.25)
```

```
    ax.zaxis.set_rotate_label(False)
    ax.set_zlabel(r' $E_{Z}$', rotation=90, labelpad=10,
                  fontsize=14)
    ax.set_zticks([-0.5, 0, 0.5, 1])
    ax.set_xlabel('cm')
    ax.set_ylabel('cm')
    ax.set_xticks(np.arange(0, 50, step=20))
    ax.set_yticks(np.arange(0, 50, step=20))
    ax.text2D(0.6, 0.7, "T = {}".format(timestep),
              transform=ax.transAxes)
    ax.xaxis.pane.fill = ax.yaxis.pane.fill = \
        ax.zaxis.pane.fill = False
    plt.gca().patch.set_facecolor('white')
    ax.text2D(-0.05, 0.8, "({})".format(label), transform=ax.
              transAxes)
    ax.dist = 11


# Plot the E field at each of the four time steps saved earlier
for subplot_num, plotting_point in enumerate(plotting_points):
    ax = fig.add_subplot(2, 2, subplot_num + 1, projection='3d')
    plot_e_field(ax, plotting_point['data_to_plot'],
                 plotting_point['num_steps'], plotting_point
                   ['label'])

fig.tight_layout()
plt.show()

# Plot Fig. 3.10
plt.rcParams['font.size'] = 12
fig = plt.figure(figsize=(8, 8))
compare_array = np.arange(-10, 10, step=1)
x_array = np.arange(-25, 25, step=1)

# The data here was generated with the Bessel function expansion
  program
compare_amp = np.array(
    [[0.3685, 0.3741, 0.3808, 0.3885, 0.3967, 0.4051, 0.4136,
      0.4220, 0.4300, 0.4376, 0.4445, 0.4507, 0.4562, 0.4610,
      0.4650, 0.4683, 0.4708, 0.4728, 0.4742, 0.4753],
     [0.3089, 0.2887, 0.2743, 0.2661, 0.2633, 0.2646, 0.2689,
      0.2754, 0.2837, 0.2932, 0.3023, 0.3090, 0.3107, 0.3048,
      0.2893, 0.2633, 0.2272, 0.1825, 0.1326, 0.0830],
     [0.2996, 0.2936, 0.2821, 0.2526, 0.2336, 0.2518, 0.2709,
      0.2539, 0.2236, 0.2479, 0.3161, 0.3622, 0.3529, 0.2909,
      0.1990, 0.1094, 0.0727, 0.0922, 0.0935, 0.0605]])
```

```
compare_phase = np.array(
    [[-212.139, -216.837, -221.089, -224.889, -228.237,
      -231.145, -233.624, -235.691, -237.360, -238.657,
      -239.561, -240.118, -240.324, -240.188, -239.713,
      -238.904, -237.761, -236.286, -234.477, -232.333],
     [-169.783, -191.511, -213.645, -235.729, -257.262,
      -277.892, -297.459, -315.879, -333.025, -348.844,
      -2.666, -14.776, -24.963, -33.243, -39.655, -44.194,
      -46.711, -46.718, -42.781, -30.063],
     [-149.660, -196.637, -240.366, -285.484, -336.898,
      -26.642, -67.400, -106.171, -153.880, -207.471,
      -246.319, -274.083, -297.293, -319.952, -345.740,
      -24.116, -94.689, -147.581, -174.128, -196.241]])


def plot_amp(ax, data, compare, freq):
    """Plot the Fourier transform amplitude at a specific
       frequency"""
    ax.plot(x_array, data, color='k', linewidth=1)
    ax.plot(compare_array, compare, 'ko', mfc='none',
            linewidth=1)
    plt.xlabel('cm')
    plt.ylabel('Amplitude')
    plt.xticks(np.arange(-5, 10, step=5))
    plt.xlim(-10, 9)
    plt.yticks(np.arange(0, 1, step=0.5))
    plt.ylim(0, 1)
    ax.text(20, 0.6, '{} MHz'.format(int(freq / 1e6)),
            horizontalalignment='center')

def plot_phase(ax, data, compare):
    """Plot the Fourier transform phase at a specific
       frequency"""
    ax.plot(x_array, data * 180 / pi, color='k', linewidth=1)
    ax.plot(compare_array, compare, 'ko', mfc='none', linewidth=1)
    plt.xlabel('cm')
    plt.ylabel('Phase (deg.)')
    plt.xticks(np.arange(-5, 10, step=5))
    plt.xlim(-10, 9)
    plt.yticks([-300, -200, -100, 0])
    plt.ylim(-360, 0)


# Plot the results of the Fourier transform at each of the
  frequencies
```

```
for m in range(number_of_frequencies):
    ax = fig.add_subplot(3, 2, m * 2 + 1)
    plot_amp(ax, amp[m], compare_amp[m], freq[m])
    ax = fig.add_subplot(3, 2, m * 2 + 2)
    plot_phase(ax, phase[m], compare_phase[m])

plt.tight_layout()
plt.show()
```

4

THREE-DIMENSIONAL SIMULATION

At last we have come to three-dimensional simulation. In actuality, three-dimensional FDTD simulation is very similar to two-dimensional simulation—only harder. It is harder because of logistical problems: we use all vector fields and each one is in three dimensions. Nonetheless, paying attention and building the programs carefully, the process is straightforward.

4.1 FREE-SPACE SIMULATION

The original FDTD paradigm was described by the *Yee cell* (Fig. 4.1) (1). Note that the E and H fields are assumed interleaved around a cell whose origin is at the locations i, j, and k. Every E field is located half a cell width from the origin in the direction of its orientation; every H field is offset half a cell in each direction except that of its orientation.

Not surprisingly, we will start with Maxwell's equations:

$$\frac{\partial \tilde{\boldsymbol{D}}}{\partial t} = \frac{1}{\sqrt{\varepsilon_0 \mu_0}} \nabla \times \boldsymbol{H}, \tag{4.1a}$$

Electromagnetic Simulation Using the FDTD Method with Python, Third Edition.
Jennifer E. Houle and Dennis M. Sullivan.

Published 2020 by John Wiley & Sons, Inc.

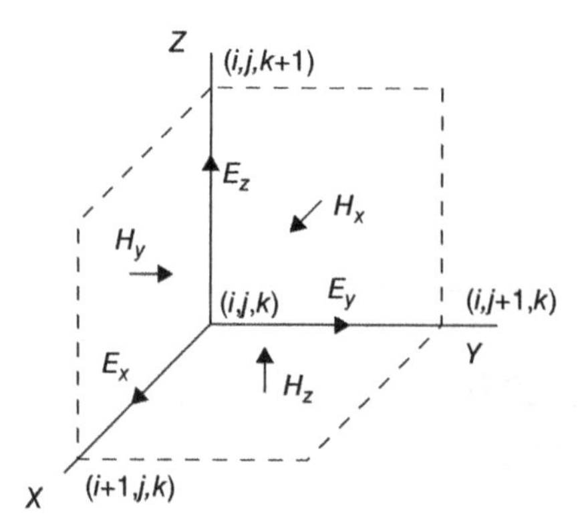

Figure 4.1 The Yee cell.

$$\widetilde{\boldsymbol{D}}(\omega) = \varepsilon_r^*(\omega)\cdot\widetilde{\boldsymbol{E}}(\omega), \tag{4.1b}$$

$$\frac{\partial \boldsymbol{H}}{\partial t} = -\frac{1}{\sqrt{\varepsilon_0\mu_0}}\nabla\times\widetilde{\boldsymbol{E}}. \tag{4.1c}$$

Once again, we drop the notation ~, but assume that we are referring to the normalized values.

Equations (4.1a) and (4.1c) produce six scalar equations:

$$\frac{\partial D_x}{\partial t} = \frac{1}{\sqrt{\varepsilon_0\mu_0}}\left(\frac{\partial H_z}{\partial y} - \frac{\partial H_y}{\partial z}\right), \tag{4.2a}$$

$$\frac{\partial D_y}{\partial t} = \frac{1}{\sqrt{\varepsilon_0\mu_0}}\left(\frac{\partial H_x}{\partial z} - \frac{\partial H_z}{\partial x}\right), \tag{4.2b}$$

$$\frac{\partial D_z}{\partial t} = \frac{1}{\sqrt{\varepsilon_0\mu_0}}\left(\frac{\partial H_y}{\partial x} - \frac{\partial H_x}{\partial y}\right), \tag{4.2c}$$

$$\frac{\partial H_x}{\partial t} = \frac{1}{\sqrt{\varepsilon_0\mu_0}}\left(\frac{\partial E_y}{\partial z} - \frac{\partial E_z}{\partial y}\right), \tag{4.2d}$$

$$\frac{\partial H_y}{\partial t} = \frac{1}{\sqrt{\varepsilon_0\mu_0}}\left(\frac{\partial E_z}{\partial x} - \frac{\partial E_x}{\partial z}\right), \tag{4.2e}$$

$$\frac{\partial H_z}{\partial t} = \frac{1}{\sqrt{\varepsilon_0\mu_0}}\left(\frac{\partial E_x}{\partial y} - \frac{\partial E_y}{\partial x}\right). \tag{4.2f}$$

The first step is to take the finite-difference approximations. We will use only Eq. (4.2c) and (4.2f) as examples:

$$D_z^{n+1/2}\left(i,j,k+\frac{1}{2}\right) = D_z^{n-1/2}\left(i,j,k+\frac{1}{2}\right)$$
$$+\frac{\Delta t}{\Delta x\cdot\sqrt{\varepsilon_0\mu_0}}\begin{bmatrix} H_y^n\left(i+\frac{1}{2},j,k+\frac{1}{2}\right)-H_y^n\left(i-\frac{1}{2},j,k+\frac{1}{2}\right) \\ -H_x^n\left(i,j+\frac{1}{2},k+\frac{1}{2}\right)+H_x^n\left(i,j-\frac{1}{2},k+\frac{1}{2}\right)\end{bmatrix}, \tag{4.3a}$$

$$H_z^{n+1}\left(i+\frac{1}{2},j+\frac{1}{2},k\right) = H_z^n\left(i+\frac{1}{2},j+\frac{1}{2},k\right)$$
$$+\frac{\Delta t}{\Delta x\cdot\sqrt{\varepsilon_0\mu_0}}\begin{bmatrix} E_x^{n+1/2}\left(i+\frac{1}{2},j+1,k\right)-E_x^{n+1/2}\left(i+\frac{1}{2},j,k\right) \\ -E_y^{n+1/2}\left(i+1,j+\frac{1}{2},k\right)+E_y^{n+1/2}\left(i,j+\frac{1}{2},k\right)\end{bmatrix}. \tag{4.3b}$$

From the difference equations, the computer equations can be written:

```
dx[i, j, k] = dx[i, j, k] + 0.5 * (
                    hz[i, j, k] - hz[i, j - 1, k] -
                    hy[i, j, k] + hy[i, j, k - 1])
dy[i, j, k] = dy[i, j, k] + 0.5 * (
                    hx[i, j, k] - hx[i, j, k - 1] -
                    hz[i, j, k] + hz[i - 1, j, k])
dz[i, j, k] = dz[i, j, k] + 0.5 * (
                    hy[i, j, k] - hy[i - 1, j, k] -
                    hx[i, j, k] + hx[i, j - 1, k])
hx[i, j, k] = hx[i, j, k] + 0.5 * (
                    ey[i, j, k + 1] - ey[i, j, k] -
                    ez[i, j + 1, k] + ez[i, j, k])
hy[i, j, k] = hy[i, j, k] + 0.5 * (
                    ez[i + 1, j, k] - ez[i, j, k] -
                    ex[i, j, k + 1] + ex[i, j, k])
hz[i, j, k] = hz[i, j, k] + 0.5 * (
                    ex[i, j + 1, k] - ex[i, j, k] -
                    ey[i + 1, j, k] + ey[i, j, k])
```

The relationship between E and D, corresponding to Eq. (4.1b), is exactly the same as the one-dimensional or two-dimensional cases, except there will be three dimensions.

The program fd3d_4_1.py at the end of this chapter is a basic three-dimensional FDTD program with a source in the middle of the problem space. This is similar to the two-dimensional program fd2d_3_1.py, except that the source is not a simple point source. In three dimensions, the E field attenuates as the square of the distance as it propagates out from the point source, so we would have trouble just seeing it. Instead, we use a dipole antenna as the source. A simple dipole antenna is illustrated in Fig. 4.2, and it consists of two metal arms. A dipole antenna functions by having current run through the arms, which results in radiation. FDTD simulates a dipole in the following manner: the metal of the arms is specified by setting the `gaz` parameters to zero in the cells corresponding to metal (Eq. (2.9) and Eq. (2.10)). This ensures that the corresponding E_z field at this point remains zero, as it would if that point were inside metal. The source is specified by setting the E_z field in the gap to a certain value. In fd3d_4_1.py, we specify a Gaussian pulse. (In a real dipole antenna, the E_z field in the gap would be the result of the current running through the metal arms.) Notice that we could have specified a current in the following manner: Ampere's circuital law says (2)

$$\int_C \boldsymbol{H} \cdot dl = \boldsymbol{I};$$

that is, the current through a surface is equal to the line integral of the H field around the surface. We could specify the current by setting the appropriate H fields around the gap as in Fig. 4.2. We can see that specifying the E field in the gap is easier.

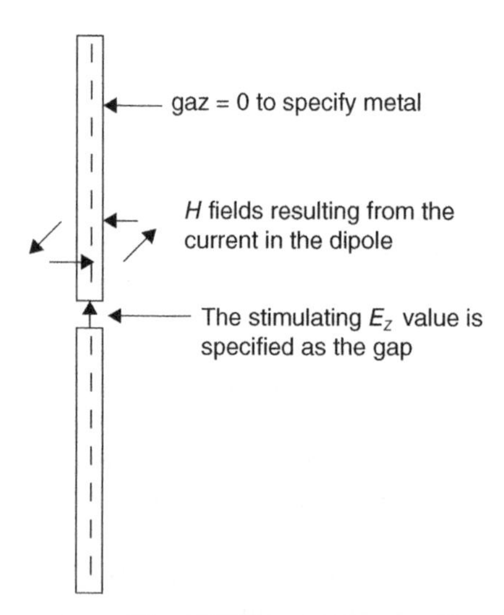

Figure 4.2 A dipole antenna. The FDTD program specifies the metal arms of the dipole by setting `gaz = 0`. The source is specified by setting the E_z field to a value in the gap.

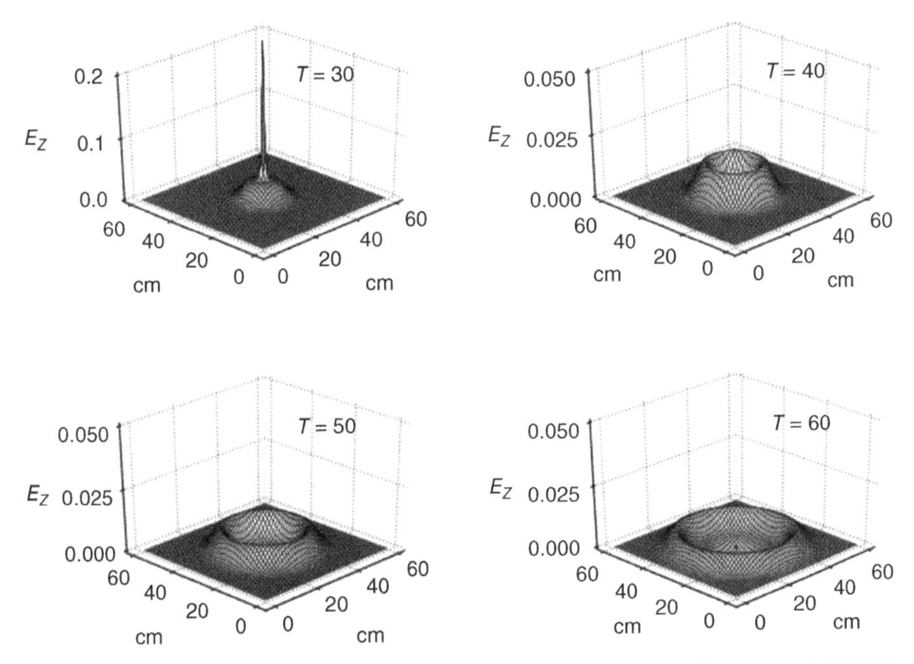

Figure 4.3 E_z field radiation from a dipole antenna in a three-dimensional FDTD program.

Figure 4.3 shows the propagation of the E_z field from the dipole in the XY plane level with the gap of the dipole. Of course, there is radiation in the Z direction as well. This illustrates a major problem in three-dimensional simulations: Unless one has unusually good graphics, visualizing three dimensions can be difficult.

PROBLEM SET 4.1

1. Get the program fd3d_4_1.py running. Duplicate the results of Fig. 4.3. (Remember, there is no PML yet.) This program introduces the Python library Numba. The large arrays used in three-dimensional simulation will cause standard Python programs to run slowly. Numba allows code to be written in Python, but with performance comparable to C. The main FDTD loop is broken into functions, each of which utilize the Numba library.

4.2 THE PML IN THREE DIMENSIONS

The development of the PML for three dimensions closely follows the two-dimensional version, but we deal with three dimensions instead of two (3). For instance, Eq. (3.23a) becomes

$$j\omega\left[1+\frac{\sigma_D(x)}{j\omega\varepsilon_0}\right]\left[1+\frac{\sigma_D(y)}{j\omega\varepsilon_0}\right]\left[1+\frac{\sigma_D(z)}{j\omega\varepsilon_0}\right]^{-1}D_z=c_0\cdot\left(\frac{\partial H_y}{\partial x}-\frac{\partial H_x}{\partial y}\right). \tag{4.4}$$

Implementing it will closely follow the two-dimensional development. Start by rewriting Eq. (4.4) as

$$\begin{aligned}j\omega\left[1+\frac{\sigma_D(x)}{j\omega\varepsilon_0}\right]\left[1+\frac{\sigma_D(y)}{j\omega\varepsilon_0}\right]D_z&=c_0\left[1+\frac{\sigma_D(z)}{j\omega\varepsilon_0}\right]\cdot\left(\frac{\partial H_y}{\partial x}-\frac{\partial H_x}{\partial y}\right)\\&=c_0\cdot\left(\frac{\partial H_y}{\partial x}-\frac{\partial H_x}{\partial y}\right)+c_0\frac{\sigma_D(z)}{\varepsilon_0}\cdot\frac{1}{j\omega}\cdot\left(\frac{\partial H_y}{\partial x}-\frac{\partial H_x}{\partial y}\right).\end{aligned} \tag{4.5}$$

We will define

$$I_{Dz}=\frac{1}{j\omega}\cdot curl\_h,$$

which is an integration when it goes to the time domain; so, Eq. (4.5) becomes

$$j\omega\left[1+\frac{\sigma_D(x)}{j\omega\varepsilon_0}\right]\left[1+\frac{\sigma_D(y)}{j\omega\varepsilon_0}\right]D_z=\frac{c_0}{\Delta x}\cdot\left[curl\_h+\frac{\sigma_D(z)}{\varepsilon_0}I_{Dz}\right].$$

Notice the Δx is placed outside the brackets and not included in *curl_h*. The implementation of this into FDTD parallels that of the two-dimensional PML, except the right side contains the integration term I_{Dz}. Therefore, following the same math we used in Chapter 3, we get

$$\begin{aligned}curl\_h=\;&H_y^n\left(i+\frac{1}{2},j,k+\frac{1}{2}\right)-H_y^n\left(i-\frac{1}{2},j,k+\frac{1}{2}\right)\\&-H_x^n\left(i,j+\frac{1}{2},k+\frac{1}{2}\right)+H_x^n\left(i,j-\frac{1}{2},k+\frac{1}{2}\right),\end{aligned} \tag{4.6a}$$

$$I_{Dz}^{n+1/2}\left(i,j,k+\frac{1}{2}\right)=I_{Dz}^{n-1/2}\left(i,j,k+\frac{1}{2}\right)+curl\_h, \tag{4.6b}$$

$$\begin{aligned}D_z^{n+1/2}\left(i,j,k+\frac{1}{2}\right)=\;&gi3(i)\cdot gj3(j)\cdot D_z^{n-1/2}\left(i,j,k+\frac{1}{2}\right)\\&+gi2(i)\cdot gj2(j)\cdot\left[0.5\cdot curl\_h+gk1(k)\cdot I_{Dz}^{n+1/2}\left(i,j,k+\frac{1}{2}\right)\right].\end{aligned} \tag{4.6c}$$

The one-dimensional g parameters are defined the same as Eq. (3.25). The computer code for Eq. (4.6) is as follows:

```
for i in range(1, ie):
    for j in range(1, je):
        for k in range(1, ke):
            curl_h = (hy[i, j, k] - hy[i - 1, j, k] -
                      hx[i, j, k] + hx[i, j - 1, k])
            idz[i, j, k] = idz[i, j, k] + curl_h
            dz[i, j, k] = gi3[i] * gj3[j] * dz[i, j, k] + \
                          gi2[i] * gj2[j] * \
                          (0.5 * curl_h + gk1[k] * idz[i, j, k])
```

The PML described in this section is a variation on the original by Berenger (3). The PML was modified to fit the FDTD formulation utilizing a separate calculation of E from D (4). This version of the PML has proven to be very effective when the background medium is air, as well as when other media such as human tissues are present and the PML are present in the other media (5). The reader should be aware that a large number of specialized PMLs have been developed. Sometimes a specialized PML is necessary for a particular medium or application (6). One of the most popular approaches used in the development of a PML is the use of stretched coordinates (7). A summary of the diverse approaches to the PML is given in Berenger (8).

PROBLEM SET 4.2

1. Add the PML to your three-dimensional program. The program fd3d_4_3.py at the end of the chapter has the PML. The program also has code used in the total/scattered field formulation described in the next section, but you should be able to pick out the code having to do with the PML. Duplicate the results of Fig. 4.4.

4.3 TOTAL/SCATTERED FIELD FORMULATION IN THREE DIMENSIONS

Generating plane waves in three dimensions is similar to doing so in two dimensions. The three-dimensional problem is illustrated in Fig. 4.5. A plane wave is generated in one plane of the three-dimensional problem space, in this case an XZ plane, at $j = ja$ and subtracted at $j = jb$. Therefore, in free space with no obstacle in the total field, we should see only E_z and H_x.

The plane wave is generated at one side and subtracted from the other by adding to $\boldsymbol{D}$ or $\boldsymbol{H}$ fields on the boundary and subtracting from $\boldsymbol{D}$ or $\boldsymbol{H}$ fields

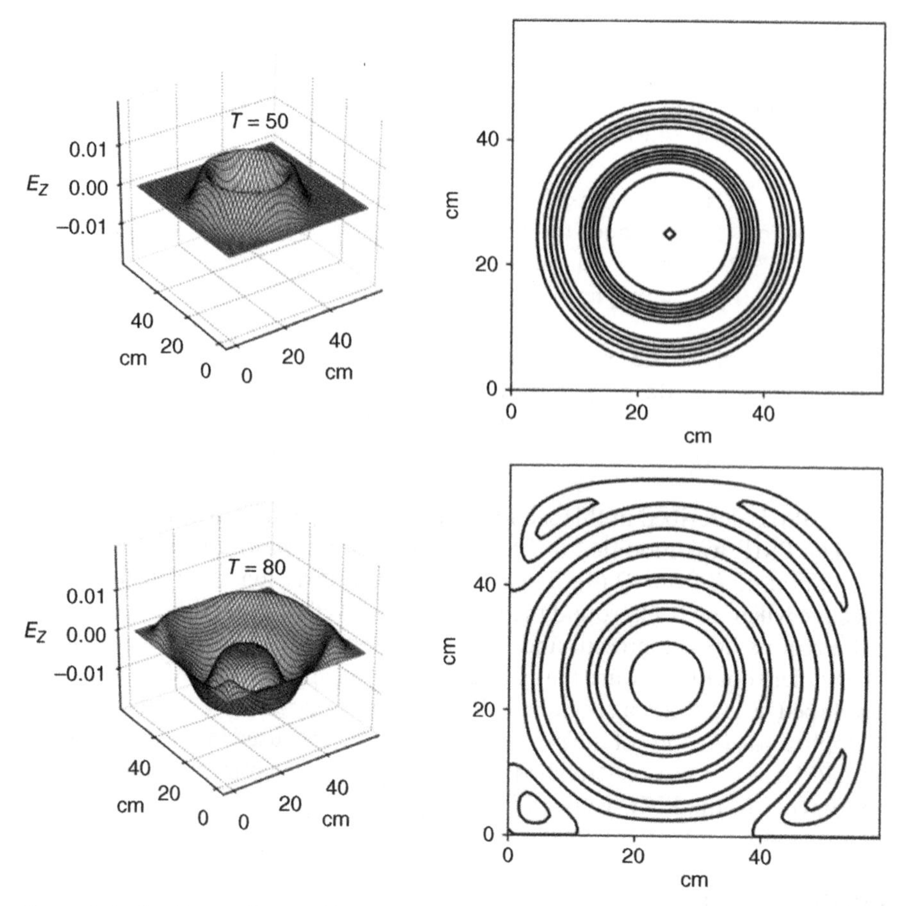

Figure 4.4 Radiation from a dipole antenna in an FDTD program with a seven-point PML. The contour diagrams on the right illustrate the fact that the fields remain concentric until they reach the PML.

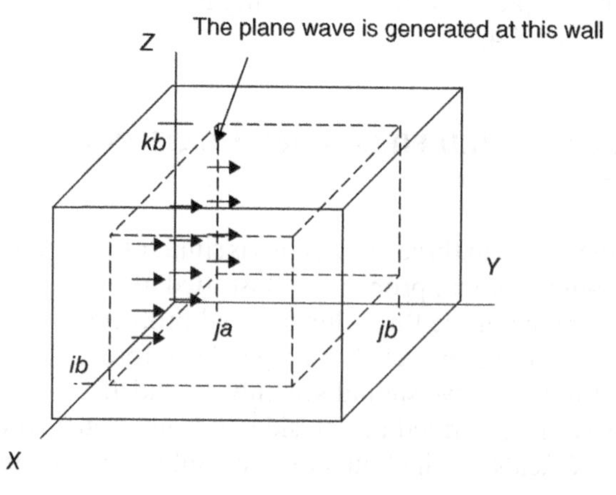

Figure 4.5 Total/scattered field in three dimensions.

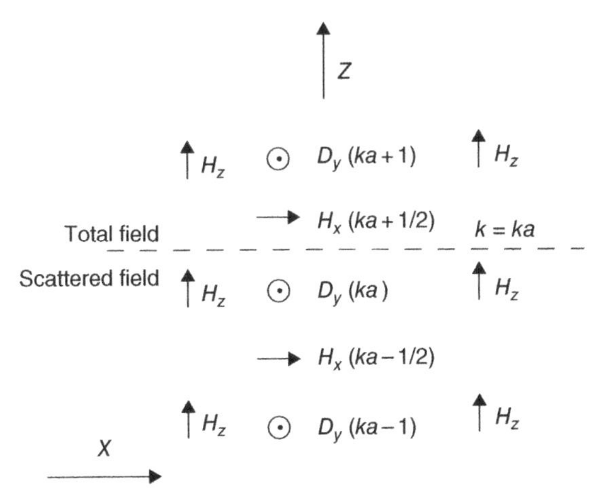

Figure 4.6 Total/scattered field boundary at $k = ka$.

that are next to the boundary. Therefore, Eq. (3.26), Eq. (3.27), and Eq. (3.28) are still used, but they are imposed on an entire plane instead of simply a line as in two dimensions. Another difference in three dimensions is the additional surfaces at $k = ka$ and $k = kb$. Figure 4.6 illustrates the $k = ka$ boundary. The calculation of $D_y\left(i, j + \frac{1}{2}, ka\right)$, which is in the scattered field, requires the values of $H_x\left(i, j + \frac{1}{2}, ka + \frac{1}{2}\right)$, which is in the total field. The difference between the two is the incident component of the H_x field. Therefore, beside equations similar to Eq. (3.26), Eq. (3.27), and Eq. (3.28), the following are necessary:

$$D_y\left(i, j + \frac{1}{2}, ka\right) = D_y\left(i, j + \frac{1}{2}, ka\right) - 0.5 \cdot H_{x\mathrm{inc}}\left(j + \frac{1}{2}\right), \tag{4.7a}$$

$$D_y\left(i, j + \frac{1}{2}, kb + 1\right) = D_y\left(i, j + \frac{1}{2}, kb + 1\right) + 0.5 \cdot H_{x\mathrm{inc}}\left(j + \frac{1}{2}\right). \tag{4.7b}$$

4.3.1 A Plane Wave Impinging on a Dielectric Sphere

Now that we have a program that generates a plane wave in three dimensions, we will begin putting objects in the problem space to see how the plane wave interacts with them. We chose a cylinder in two dimensions because we had an analytic solution we could use to check the accuracy of our calculation via a Bessel function expansion. It turns out that the interaction of a plane wave with a dielectric sphere can be determined by an expansion of the modified Bessel functions (9).

Specifying a sphere in three dimensions is similar to specifying a cylinder in two dimensions. Again, the major difference is that it must be done for all three electric fields. The following code specifies the parameters needed for the E_z field calculation:

```
for i in range(ia, ib + 1):
    for j in range(ja, jb + 1):
        for k in range(ka, kb + 1):
            eps = epsilon[0]
            cond = sigma[0]
            xdist = ic - i
            ydist = jc - j
            zdist = kc - k - 0.5
            dist = sqrt(xdist ** 2 + ydist ** 2 + zdist ** 2)
            if dist <= radius:
                eps = epsilon[1]
                cond = sigma[1]
            gaz[i, j, k] = 1 / (eps + (cond * dt / epsz))
            gbz[i, j, k] = cond * dt / epsz
```

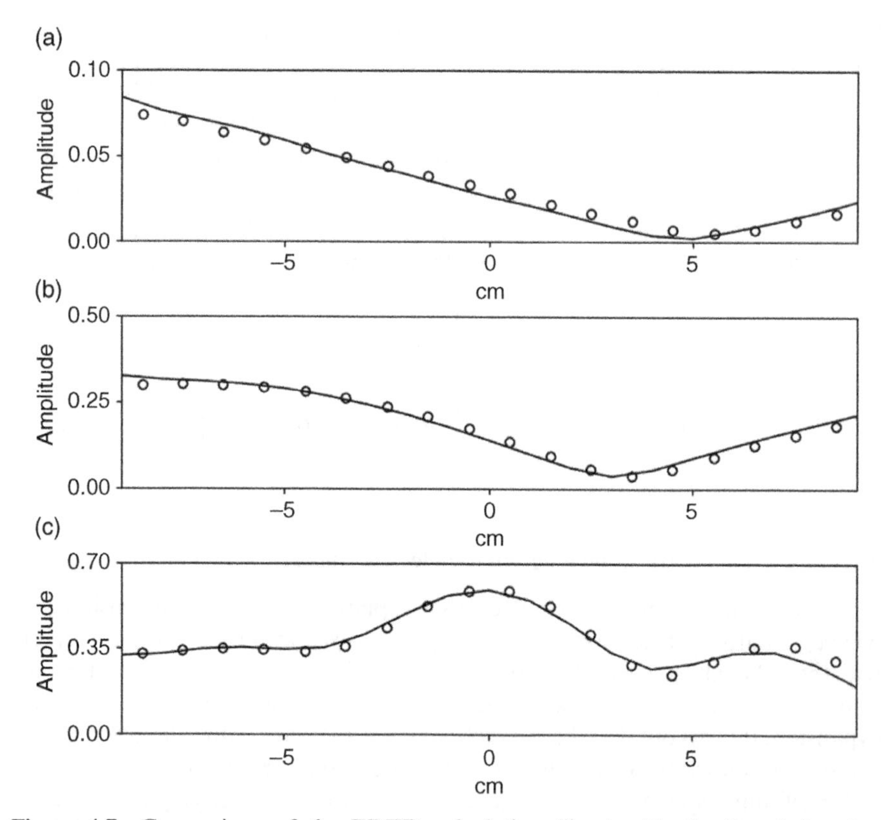

Figure 4.7 Comparison of the FDTD calculation (lines) with the Bessel function expansion calculation (circles) along the main axis of a dielectric sphere, 20 cm in diameter, with $\varepsilon_r = 30$ and $\sigma = 0.3$. The program for the FDTD calculation used the simple "in-or-out" strategy to determine the parameters. (a) 50, (b), 200, and (c) 500 MHz.

Besides the obvious differences such as three loops instead of two, another difference can be distinguished. The parameter `zdist` calculates the distance to the point as if it were half a cell further in the z direction. That is because it is! As shown in Fig. 4.1, each E field is assumed offset from i, j, and k by half a cell in its own direction.

In Fig. 4.7, we show a comparison between FDTD results and the Bessel function expansion results for a plane wave incident on a dielectric sphere. The sphere has a diameter of 20 cm, a dielectric constant of 30, and a conductivity of 0.3. The FDTD program is 40 × 40 × 40 cells and uses a seven-point PML. Apparently, the "in-or-out method" of determining the parameters such as `gaz` and `gbz` is not as forgiving as it was in two dimensions.

We can do an averaging over subcells to improve efficiency. We might conclude that because this means averaging subcells in a plane in two dimensions, we would have to average subcubes in three dimensions. It turns out that this is not the case. Figure 4.8 illustrates the calculation of E_z, which uses the surrounding H_x and H_y values. As this calculation is confined to a plane, we may think of the calculation of E_z as being a two-dimensional problem. Therefore, in determining the parameters `gaz` and `gbz`, we will consider how much of the area of the plane containing H_x and H_y exists in each of the different media.

The following code averages the properties over nine subcells:

```
for i in range(ia, ib + 1):
    for j in range(ja, jb + 1):
        for k in range(ka, kb + 1):
            eps = epsilon[0]
            cond = sigma[0]
            for jj in range(-1, 2):
                for ii in range(-1, 2):
                    xdist = (ic - i) + 1 / 3 * ii
```

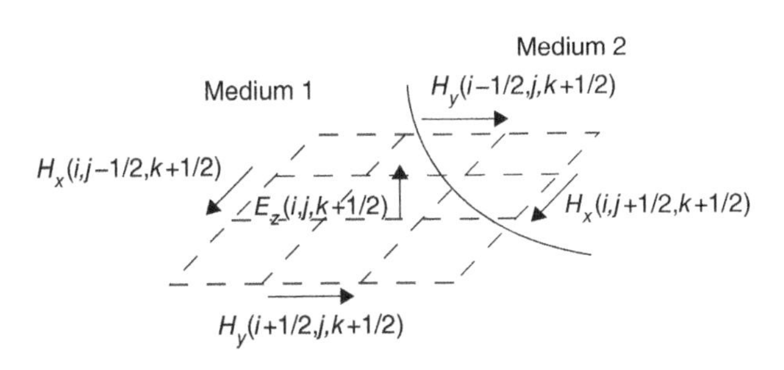

Figure 4.8 E_z is calculated by the surrounding H_x and H_y values. The parameters used to calculate E_z are determined by the percentage of subcells that lie in each medium. In this example, six subcells are in medium 1 and three subcells are in medium 2.

```
                ydist = (jc - j) + 1 / 3 * jj
                zdist = (kc - k) - 0.5
                dist = sqrt(xdist ** 2 +
                            ydist ** 2 + zdist ** 2)
                if dist <= radius:
                    eps = eps + (1 / 3 ** 2) * (epsilon[1] -
                                                epsilon[0])
                    cond = cond + (1 / 3 ** 2) * (sigma[1])
        gaz[i, j, k] = 1 / (eps + (cond * dt / epsz))
        gbz[i, j, k] = cond * dt / epsz
```

Figure 4.9 repeats the FDTD/Bessel comparison, but with the averaged parameter values. Clearly, it results in an improvement.

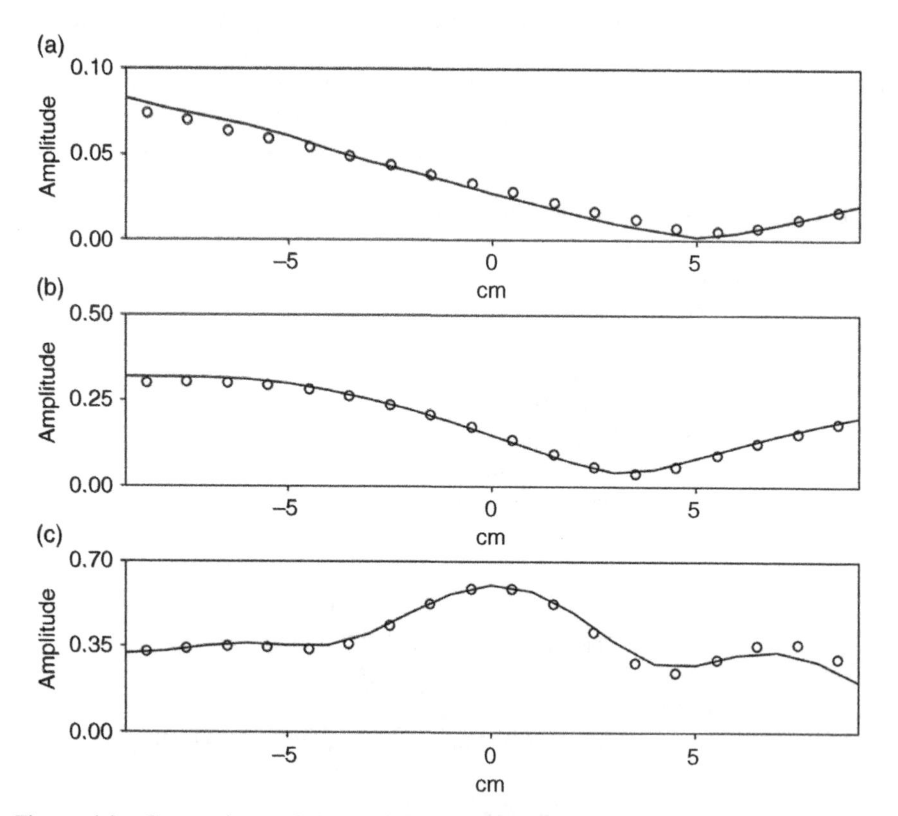

Figure 4.9 Comparison of the FDTD calculation (lines) with the Bessel function expansion calculation (circles) along the main axis of a dielectric sphere, 20 cm in diameter, with $\varepsilon_r = 30$ and $\sigma = 0.3$. The program used for the FDTD calculation averaged over nine subcells within each cell to determine the parameters. (a) 50, (b) 200, and (c) 500 MHz.

PROBLEM SET 4.3

1. Add the plane wave propagation to your three-dimensional FDTD program. Make sure that it can propagate a wave through and out the other end. (See the program fd3d_4_3.py.)
2. The program fd3d_4_3.py creates the spheres by the "in-or-out." Get this program running and duplicate the results of Fig. 4.7.
3. Use the averaging technique to get a more accurate calculation of the parameters and duplicate the results of Fig. 4.9.

REFERENCES

1. K. S. Yee, Numerical solution of initial boundary value problems involving Maxwell's equations in isotropic media, *IEEE Trans. Antennas Propag.*, vol. AP-17, 1996, pp. 585–589.
2. D. K. Cheng, *Field and Wave Electromagnetics*, Menlo Park, CA: Addison-Wesley, 1992.
3. J. P. Berenger, A perfectly matched layer for the absorption of electromagnetic waves, *J. Comput. Phys.*, vol. 114, 1994, pp. 185–200.
4. D. M. Sullivan, An unsplit step 3D PML for use with the FDTD Method, *IEEE Microw. Guid. Wave Lett.*, vol. 7, Feb. 1997, pp. 184–186.
5. D. M. Sullivan, P. Wust, and J. Nadobny, Accurate FDTD simulation of RF coils for MRI using the thin-rod approximation, *IEEE Trans. Antennas Propag.*, vol. 58, Aug. 2003, pp. 1780–1796.
6. D. M. Sullivan, Y. Xia, and D. Butherus, A perfectly matched layer for lossy media at extremely low frequencies, *IEEE Antennas Wirel. Propag. Lett.*, vol. 8, 2009, pp. 1080–1083.
7. W. C. Chew and W. H. Weedon, A 3D perfectly matched medium form modified Maxwell's equations with stretched coordinates, *Microwave Opt. Technol. Lett.*, vol. 3, 1998, pp. 559–605.
8. J. P Berenger, *Perfectly Matched Layer (PML) for Computational Electromagnetics*, Morgan and Claypool Publishers, 2007.
9. R. Harrington, *Time-Harmonic Electromagnetic Fields*, New York: McGraw-Hill, 1961.


```
""" fd2d_4_1.py: 3D FDTD

Dipole in free space
"""

import numpy as np
from math import exp
from matplotlib import pyplot as plt
from mpl_toolkits.mplot3d.axes3d import Axes3D
import numba
```

```
ie = 60
je = 60
ke = 60
ic = int(ie / 2)
jc = int(je / 2)
kc = int(ke / 2)

ex = np.zeros((ie, je, ke))
ey = np.zeros((ie, je, ke))
ez = np.zeros((ie, je, ke))
dx = np.zeros((ie, je, ke))
dy = np.zeros((ie, je, ke))
dz = np.zeros((ie, je, ke))
hx = np.zeros((ie, je, ke))
hy = np.zeros((ie, je, ke))
hz = np.zeros((ie, je, ke))
gax = np.ones((ie, je, ke))
gay = np.ones((ie, je, ke))
gaz = np.ones((ie, je, ke))

ddx = 0.01  # Cell size
dt = ddx / 6e8  # Time step size
epsz = 8.854e-12

# Specify the dipole
gaz[ic, jc, kc - 10:kc + 10] = 0
gaz[ic, jc, kc] = 1

# Pulse Parameters
t0 = 20
spread = 6

nsteps = 60

# At each of these four time steps, save the E field data
# for plotting at the end
plotting_points = [
    {'num_steps': 30, 'data_to_plot': None, 'z_scale': 0.20},
    {'num_steps': 40, 'data_to_plot': None, 'z_scale': 0.05},
    {'num_steps': 50, 'data_to_plot': None, 'z_scale': 0.05},
    {'num_steps': 60, 'data_to_plot': None, 'z_scale': 0.05},
]
```

```
# Functions for Main FDTD Loop

@numba.jit(nopython=True)
def calculate_d_fields(ie, je, ke, dx, dy, dz, hx, hy, hz):
    """ Calculate the Dx, Dy, and Dz fields """
    for i in range(1, ie):
        for j in range(1, je):
            for k in range(1, ke):
                dx[i, j, k] = dx[i, j, k] + 0.5 * (
                    hz[i, j, k] - hz[i, j - 1, k] -
                    hy[i, j, k] + hy[i, j, k - 1])

    for i in range(1, ie):
        for j in range(1, je):
            for k in range(1, ke):
                dy[i, j, k] = dy[i, j, k] + 0.5 * (
                    hx[i, j, k] - hx[i, j, k - 1] -
                    hz[i, j, k] + hz[i - 1, j, k])

    for i in range(1, ie):
        for j in range(1, je):
            for k in range(1, ke):
                dz[i, j, k] = dz[i, j, k] + 0.5 * (
                    hy[i, j, k] - hy[i - 1, j, k] -
                    hx[i, j, k] + hx[i, j - 1, k])

    return dx, dy, dz


@numba.jit(nopython=True)
def calculate_e_fields(ie, je, ke, dx, dy, dz,
                       gax, gay, gaz, ex, ey, ez):
    """ Calculate the E field from the D field """
    for i in range(0, ie):
        for j in range(0, je):
            for k in range(0, ke):
                ex[i, j, k] = gax[i, j, k] * dx[i, j, k]
                ey[i, j, k] = gay[i, j, k] * dy[i, j, k]
                ez[i, j, k] = gaz[i, j, k] * dz[i, j, k]

    return ex, ey, ez
```

```
@numba.jit(nopython=True)
def calculate_h_fields(ie, je, ke, hx, hy, hz, ex, ey, ez):
    """ Calculate the Hx, Hy, and Hz fields """
    for i in range(0, ie):
        for j in range(0, je - 1):
            for k in range(0, ke - 1):
                hx[i, j, k] = hx[i, j, k] + 0.5 * (
                    ey[i, j, k + 1] - ey[i, j, k] -
                    ez[i, j + 1, k] + ez[i, j, k])

    for i in range(0, ie - 1):
        for j in range(0, je):
            for k in range(0, ke - 1):
                hy[i, j, k] = hy[i, j, k] + 0.5 * (
                    ez[i + 1, j, k] - ez[i, j, k] -
                    ex[i, j, k + 1] + ex[i, j, k])

    for i in range(0, ie - 1):
        for j in range(0, je - 1):
            for k in range(0, ke):
                hz[i, j, k] = hz[i, j, k] + 0.5 * (
                    ex[i, j + 1, k] - ex[i, j, k] -
                    ey[i + 1, j, k] + ey[i, j, k])

    return hx, hy, hz


# Main FDTD Loop
for time_step in range(1, nsteps + 1):

    # Calculate the D Fields
    dx, dy, dz = calculate_d_fields(ie, je, ke, dx, dy, dz, hx, hy, hz)

    # Add the source at the gap
    pulse = exp(-0.5 * ((t0 - time_step) / spread) ** 2)
    dz[ic, jc, kc] = pulse

    # Calculate the E field from the D field
    ex, ey, ez = calculate_e_fields(ie, je, ke, dx, dy, dz,
                                   gax, gay, gaz, ex, ey, ez)

    # Calculate the H fields
    hx, hy, hz = calculate_h_fields(ie, je, ke, hx, hy, hz, ex, ey, ez)
```

```
    # Save data at certain points for later plotting
    for plotting_point in plotting_points:
        if time_step == plotting_point['num_steps']:
            plotting_point['data_to_plot'] = np.copy(ez)

# Plot Fig. 4.3
plt.rcParams['font.size'] = 12
plt.rcParams['grid.color'] = 'gray'
plt.rcParams['grid.linestyle'] = 'dotted'
fig = plt.figure(figsize=(8, 6))

X, Y = np.meshgrid(range(je), range(ie))


def plot_e_field(ax, data, timestep, scale):
    """3d Plot of E field at a single time step"""
    ax.set_zlim(0, scale)
    ax.view_init(elev=30., azim=-135)
    ax.plot_surface(X, Y, data[:, :, kc], rstride=1, cstride=1,
                    color='white', edgecolor='black', linewidth=0.25)
    ax.zaxis.set_rotate_label(False)
    ax.set_zlabel(r' $E_{Z}$', rotation=90, labelpad=10, fontsize=14)
    ax.set_zticks([0, scale / 2, scale])
    ax.set_xlabel('cm')
    ax.set_ylabel('cm')
    ax.set_xticks(np.arange(0, 61, step=20))
    ax.set_yticks(np.arange(0, 61, step=20))
    ax.text2D(0.6, 0.7, "T = {}".format(timestep),
              transform=ax.transAxes)
    ax.xaxis.pane.fill = ax.yaxis.pane.fill = \
        ax.zaxis.pane.fill = False
    plt.gca().patch.set_facecolor('white')
    ax.dist = 11


# Plot the E field at each of the four time steps saved earlier
for subplot_num, plotting_point in enumerate(plotting_points):
    ax = fig.add_subplot(2, 2, subplot_num + 1, projection='3d')
    plot_e_field(ax, plotting_point['data_to_plot'],
                 plotting_point['num_steps'],
                   plotting_point['z_scale'])

plt.subplots_adjust(bottom=0.05, left=0.05)
plt.show()
```

```
""" fd3d_4_3.py: 3d FDTD

Chapter 4 Section 3
3D FDTD simulation of a plane wave on a dielectric sphere
"""

from math import exp, sqrt, cos, sin

import numba
import numpy as np
from matplotlib import pyplot as plt


def calculate_pml_parameters(npml, ie, je, ke):
    """ Calculate and return the PML parameters """
    gi1 = np.zeros(ie)
    gi2 = np.ones(ie)
    gi3 = np.ones(ie)
    fi1 = np.zeros(ie)
    fi2 = np.ones(ie)
    fi3 = np.ones(ie)

    gj1 = np.zeros(je)
    gj2 = np.ones(je)
    gj3 = np.ones(je)
    fj1 = np.zeros(je)
    fj2 = np.ones(je)
    fj3 = np.ones(je)

    gk1 = np.zeros(ke)
    gk2 = np.ones(ke)
    gk3 = np.ones(ke)
    fk1 = np.zeros(ke)
    fk2 = np.ones(ke)
    fk3 = np.ones(ke)

    for n in range(npml):
        xxn = (npml - n) / npml
        xn = 0.33 * (xxn ** 3)

        fi1[n] = xn
        fi1[ie - n - 1] = xn
        gi2[n] = 1 / (1 + xn)
        gi2[ie - 1 - n] = 1 / (1 + xn)
```

```
        gi3[n] = (1 - xn) / (1 + xn)
        gi3[ie - 1 - n] = (1 - xn) / (1 + xn)

        fj1[n] = xn
        fj1[je - n - 1] = xn
        gj2[n] = 1 / (1 + xn)
        gj2[je - 1 - n] = 1 / (1 + xn)
        gj3[n] = (1 - xn) / (1 + xn)
        gj3[je - 1 - n] = (1 - xn) / (1 + xn)

        fk1[n] = xn
        fk1[ke - n - 1] = xn
        gk2[n] = 1 / (1 + xn)
        gk2[ke - 1 - n] = 1 / (1 + xn)
        gk3[n] = (1 - xn) / (1 + xn)
        gk3[ke - 1 - n] = (1 - xn) / (1 + xn)

        xxn = (npml - n - 0.5) / npml
        xn = 0.33 * (xxn ** 3)

        gi1[n] = xn
        gi1[ie - 1 - n] = xn
        fi2[n] = 1 / (1 + xn)
        fi2[ie - 1 - n] = 1 / (1 + xn)
        fi3[n] = (1 - xn) / (1 + xn)
        fi3[ie - 1 - n] = (1 - xn) / (1 + xn)

        gj1[n] = xn
        gj1[je - 1 - n] = xn
        fj2[n] = 1 / (1 + xn)
        fj2[je - 1 - n] = 1 / (1 + xn)
        fj3[n] = (1 - xn) / (1 + xn)
        fj3[je - 1 - n] = (1 - xn) / (1 + xn)

        gk1[n] = xn
        gk1[ke - 1 - n] = xn
        fk2[n] = 1 / (1 + xn)
        fk2[ke - 1 - n] = 1 / (1 + xn)
        fk3[n] = (1 - xn) / (1 + xn)
        fk3[ke - 1 - n] = (1 - xn) / (1 + xn)

    return gi1, gi2, gi3, fi1, fi2, fi3, gj1, gj2, gj3, fj1, fj2, fj3, \
        gk1, gk2, gk3, fk1, fk2, fk3
```

```
@numba.jit(nopython=True)
def calculate_dx_field(ie, je, ke, dx, idx, hy, hz,
                       gj3, gk3, gj2, gk2, gi1):
    """ Calculate the Dx Field """
    for i in range(1, ie):
        for j in range(1, je):
            for k in range(1, ke):
                curl_h = (hz[i, j, k] - hz[i, j - 1, k] -
                          hy[i, j, k] + hy[i, j, k - 1])
                idx[i, j, k] = idx[i, j, k] + curl_h
                dx[i, j, k] = gj3[j] * gk3[k] * dx[i, j, k] + \
                    gj2[j] * gk2[k] * \
                    (0.5 * curl_h + gi1[i] * idx[i, j, k])
    return dx, idx


@numba.jit(nopython=True)
def calculate_dy_field(ie, je, ke, dy, idy, hx, hz,
                       gi3, gk3, gi2, gk2, gj1):
    """ Calculate the Dy Field """
    for i in range(1, ie):
        for j in range(1, je):
            for k in range(1, ke):
                curl_h = (hx[i, j, k] - hx[i, j, k - 1] -
                          hz[i, j, k] + hz[i - 1, j, k])
                idy[i, j, k] = idy[i, j, k] + curl_h
                dy[i, j, k] = gi3[i] * gk3[k] * dy[i, j, k] + \
                    gi2[i] * gk2[k] * \
                    (0.5 * curl_h + gj1[j] * idy[i, j, k])
    return dy, idy


@numba.jit(nopython=True)
def calculate_dz_field(ie, je, ke, dz, idz, hx, hy,
                       gi3, gj3, gi2, gj2, gk1):
    """ Calculate the Dz Field """
    for i in range(1, ie):
        for j in range(1, je):
            for k in range(1, ke):
                curl_h = (hy[i, j, k] - hy[i - 1, j, k] -
                          hx[i, j, k] + hx[i, j - 1, k])
```

```
                idz[i, j, k] = idz[i, j, k] + curl_h
                dz[i, j, k] = gi3[i] * gj3[j] * dz[i, j, k] + \
                              gi2[i] * gj2[j] * \
                              (0.5 * curl_h + gk1[k] * idz[i, j, k])
    return dz, idz


@numba.jit(nopython=True)
def calculate_inc_dy_field(ia, ib, ja, jb, ka, kb, dy, hx_inc):
    """ Calculate the incident Dy Field """
    for i in range(ia, ib + 1):
        for j in range(ja, jb + 1):
            dy[i, j, ka] = dy[i, j, ka] - 0.5 * hx_inc[j]
            dy[i, j, kb + 1] = dy[i, j, kb + 1] + 0.5 * hx_inc[j]
    return dy


@numba.jit(nopython=True)
def calculate_inc_dz_field(ia, ib, ja, jb, ka, kb, dz, hx_inc):
    """ Calculate the incident Dz Field"""
    for i in range(ia, ib + 1):
        for k in range(ka, kb + 1):
            dz[i, ja, k] = dz[i, ja, k] + 0.5 * hx_inc[ja - 1]
            dz[i, jb, k] = dz[i, jb, k] - 0.5 * hx_inc[jb]

    return dz


@numba.jit(nopython=True)
def calculate_e_fields(ie, je, ke, dx, dy, dz, gax, gay, gaz,
                       gbx, gby, gbz, ex, ey, ez, ix, iy, iz):
    """ Calculate the E field from the D field"""
    for i in range(0, ie):
        for j in range(0, je):
            for k in range(0, ke):
                ex[i, j, k] = gax[i, j, k] * (dx[i, j, k] - ix[i, j, k])
                ix[i, j, k] = ix[i, j, k] + gbx[i, j, k] * ex[i, j, k]
                ey[i, j, k] = gay[i, j, k] * (dy[i, j, k] - iy[i, j, k])
                iy[i, j, k] = iy[i, j, k] + gby[i, j, k] * ey[i, j, k]
                ez[i, j, k] = gaz[i, j, k] * (dz[i, j, k] - iz[i, j, k])
                iz[i, j, k] = iz[i, j, k] + gbz[i, j, k] * ez[i, j, k]

    return ex, ey, ez, ix, iy, iz
```

```
@numba.jit(nopython=True)
def calculate_fourier_transform_ex(ie, je, number_of_frequencies,
                                   real_pt, imag_pt, ez, arg,
                                   time_step, kc):
    """ Calculate the Fourier transform of Ex"""
    for i in range(0, ie):
        for j in range(0, je):
            for m in range(0, number_of_frequencies):
                real_pt[m, i, j] = real_pt[m, i, j] + \
                                   cos(arg[m] * time_step)
                                     * ez[i, j, kc]
                imag_pt[m, i, j] = imag_pt[m, i, j] - \
                                   sin(arg[m] * time_step)
                                     * ez[i, j, kc]

    return real_pt, imag_pt


@numba.jit(nopython=True)
def calculate_hx_field(ie, je, ke, hx, ihx, ey, ez,
                       fi1, fj2, fk2, fj3, fk3):
    """ Calculate the Hx field"""
    for i in range(0, ie):
        for j in range(0, je - 1):
            for k in range(0, ke - 1):
                curl_e = (ey[i, j, k + 1] - ey[i, j, k] -
                          ez[i, j + 1, k] + ez[i, j, k])
                ihx[i, j, k] = ihx[i, j, k] + curl_e
                hx[i, j, k] = fj3[j] * fk3[k] * hx[i, j, k] + \
                              fj2[j] * fk2[k] * 0.5 * \
                              (curl_e + fi1[i] * ihx[i, j, k])
    return hx, ihx


@numba.jit(nopython=True)
def calculate_hy_field(ie, je, ke, hy, ihy, ex, ez,
                       fj1, fi2, fk2, fi3, fk3):
    """ Calculate the Hy field"""
    for i in range(0, ie - 1):
        for j in range(0, je):
            for k in range(0, ke - 1):
                curl_e = (ez[i + 1, j, k] - ez[i, j, k] -
                          ex[i, j, k + 1] + ex[i, j, k])
```

```
                ihy[i, j, k] = ihy[i, j, k] + curl_e
                hy[i, j, k] = fi3[i] * fk3[k] * hy[i, j, k] + \
                              fi2[i] * fk2[k] * 0.5 * \
                              (curl_e + fj1[j] * ihy[i, j, k])
    return hy, ihy


@numba.jit(nopython=True)
def calculate_hz_field(ie, je, ke, hz, ihz, ex, ey,
                       fk1, fi2, fj2, fi3, fj3):
    """ Calculate the Hz field"""
    for i in range(0, ie - 1):
        for j in range(0, je - 1):
            for k in range(0, ke):
                curl_e = (ex[i, j + 1, k] - ex[i, j, k] -
                          ey[i + 1, j, k] + ey[i, j, k])
                ihz[i, j, k] = ihz[i, j, k] + curl_e
                hz[i, j, k] = fi3[i] * fj3[j] * hz[i, j, k] + \
                              fi2[i] * fj2[j] * 0.5 * \
                              (curl_e + fk1[k] * ihz[i, j, k])
    return hz, ihz


@numba.jit(nopython=True)
def calculate_hx_inc(je, hx_inc, ez_inc):
    """ Calculate incident Hx field"""
    for j in range(0, je - 1):
        hx_inc[j] = hx_inc[j] + 0.5 * (ez_inc[j] - ez_inc[j + 1])

    return hx_inc


@numba.jit(nopython=True)
def calculate_hx_with_incident_field(ia, ib, ja, jb, ka, kb, hx,
  ez_inc):
    """ Calculate Hx with incident ez"""
    for i in range(ia, ib + 1):
        for k in range(ka, kb + 1):
            hx[i, ja - 1, k] = hx[i, ja - 1, k] + 0.5 * ez_inc[ja]
            hx[i, jb, k] = hx[i, jb, k] - 0.5 * ez_inc[jb]

    return hx
```

```
@numba.jit(nopython=True)
def calculate_hy_with_incident_field(ia, ib, ja, jb, ka, kb, hy, ez_inc):
    """ Calculate Hy with incident ez"""
    for j in range(ja, jb + 1):
        for k in range(ka, kb + 1):
            hy[ia - 1, j, k] = hy[ia - 1, j, k] - 0.5 * ez_inc[j]
            hy[ib, j, k] = hy[ib, j, k] + 0.5 * ez_inc[j]

    return hy


ie = 40
je = 40
ke = 40
ic = int(ie / 2)
jc = int(je / 2)
kc = int(ke / 2)
ia = 7
ja = 7
ka = 7
ib = ie - ia - 1
jb = je - ja - 1
kb = ke - ka - 1

ex = np.zeros((ie, je, ke))
ey = np.zeros((ie, je, ke))
ez = np.zeros((ie, je, ke))
ix = np.zeros((ie, je, ke))
iy = np.zeros((ie, je, ke))
iz = np.zeros((ie, je, ke))
dx = np.zeros((ie, je, ke))
dy = np.zeros((ie, je, ke))
dz = np.zeros((ie, je, ke))
idx = np.zeros((ie, je, ke))
idy = np.zeros((ie, je, ke))
idz = np.zeros((ie, je, ke))
hx = np.zeros((ie, je, ke))
hy = np.zeros((ie, je, ke))
hz = np.zeros((ie, je, ke))
ihx = np.zeros((ie, je, ke))
ihy = np.zeros((ie, je, ke))
ihz = np.zeros((ie, je, ke))
```

```
gax = np.ones((ie, je, ke))
gay = np.ones((ie, je, ke))
gaz = np.ones((ie, je, ke))
gbx = np.zeros((ie, je, ke))
gby = np.zeros((ie, je, ke))
gbz = np.zeros((ie, je, ke))
hx_inc = np.zeros(je)
ez_inc = np.zeros(je)

ddx = 0.01  # Cell size
dt = ddx / 6e8  # Time step size
epsz = 8.854e-12

number_of_frequencies = 3
freq = np.array((50e6, 200e6, 500e6))
arg = 2 * np.pi * freq * dt
real_in = np.zeros(number_of_frequencies)
imag_in = np.zeros(number_of_frequencies)
real_pt = np.zeros((number_of_frequencies, ie, je, ke))
imag_pt = np.zeros((number_of_frequencies, ie, je, ke))
amp = np.zeros((number_of_frequencies, je))

# Specify the dielectric sphere
epsilon = np.ones(2)
sigma = np.zeros(2)
epsilon[1] = 30
sigma[1] = 0.3
radius = 10

for i in range(ia, ib + 1):
    for j in range(ja, jb + 1):
        for k in range(ka, kb + 1):
            eps = epsilon[0]
            cond = sigma[0]
            xdist = ic - i - 0.5
            ydist = jc - j
            zdist = kc - k
            dist = sqrt(xdist ** 2 + ydist ** 2 + zdist ** 2)
            if dist <= radius:
                eps = epsilon[1]
                cond = sigma[1]
            gax[i, j, k] = 1 / (eps + (cond * dt / epsz))
            gbx[i, j, k] = cond * dt / epsz
```

```
for i in range(ia, ib + 1):
    for j in range(ja, jb + 1):
        for k in range(ka, kb + 1):
            eps = epsilon[0]
            cond = sigma[0]
            xdist = ic - i
            ydist = jc - j - 0.5
            zdist = kc - k
            dist = sqrt(xdist ** 2 + ydist ** 2 + zdist ** 2)
            if dist <= radius:
                eps = epsilon[1]
                cond = sigma[1]
            gay[i, j, k] = 1 / (eps + (cond * dt / epsz))
            gby[i, j, k] = cond * dt / epsz


for i in range(ia, ib + 1):
    for j in range(ja, jb + 1):
        for k in range(ka, kb + 1):
            eps = epsilon[0]
            cond = sigma[0]
            xdist = ic - i
            ydist = jc - j
            zdist = kc - k - 0.5
            dist = sqrt(xdist ** 2 + ydist ** 2 + zdist ** 2)
            if dist <= radius:
                eps = epsilon[1]
                cond = sigma[1]
            gaz[i, j, k] = 1 / (eps + (cond * dt / epsz))
            gbz[i, j, k] = cond * dt / epsz


# Pulse Parameters
t0 = 20
spread = 8


# Calculate the PML parameters
npml = 8
gi1, gi2, gi3, fi1, fi2, fi3, gj1, gj2, gj3, fj1, fj2, fj3, \
gk1, gk2, gk3, fk1, fk2, fk3 = calculate_pml_parameters(npml, ie, je, ke)


boundary_low = [0, 0]
boundary_high = [0, 0]

nsteps = 500
```

```
# Main FDTD Loop
for time_step in range(1, nsteps + 1):

    # Calculate the incident buffer
    for j in range(1, je - 1):
        ez_inc[j] = ez_inc[j] + 0.5 * (hx_inc[j - 1] - hx_inc[j])

    # Fourier transform of the incident field
    for m in range(number_of_frequencies):
        real_in[m] = real_in[m] + cos(arg[m] * time_step) * ez_inc[ja - 1]
        imag_in[m] = imag_in[m] - sin(arg[m] * time_step) * ez_inc[ja - 1]

    # Absorbing Boundary Conditions
    ez_inc[0] = boundary_low.pop(0)
    boundary_low.append(ez_inc[1])

    ez_inc[je - 1] = boundary_high.pop(0)
    boundary_high.append(ez_inc[je - 2])

    # Calculate the D Fields
    dx, idx = calculate_dx_field(ie, je, ke, dx, idx, hy, hz,
                                 gj3, gk3, gj2, gk2, gi1)
    dy, idy = calculate_dy_field(ie, je, ke, dy, idy, hx, hz,
                                 gi3, gk3, gi2, gk2, gj1)
    dz, idz = calculate_dz_field(ie, je, ke, dz, idz, hx, hy,
                                 gi3, gj3, gi2, gj2, gk1)

    # Add the source at the gap
    pulse = exp(-0.5 * ((t0 - time_step) / spread) ** 2)
    ez_inc[3] = pulse

    dy = calculate_inc_dy_field(ia, ib, ja, jb, ka, kb, dy, hx_inc)
    dz = calculate_inc_dz_field(ia, ib, ja, jb, ka, kb, dz, hx_inc)

    # Calculate the E field from the D field
    ex, ey, ez, ix, iy, iz = calculate_e_fields(ie, je, ke, dx, dy, dz,
                                                gax, gay, gaz,
                                                gbx, gby, gbz,
                                                ex, ey, ez, ix, iy, iz)

    # Calculate the Fourier transform of Ex
    real_pt, imag_pt = \
      calculate_fourier_transform_ex(ie, je, number_of_frequencies,
                                     real_pt, imag_pt,
                                     ez, arg, time_step, kc)
```

```
    # Calculate the H fields
    hx_inc = calculate_hx_inc(je, hx_inc, ez_inc)
    hx, ihx = calculate_hx_field(ie, je, ke, hx, ihx, ey, ez,
                                 fi1, fj2, fk2, fj3, fk3)
    hx = calculate_hx_with_incident_field(ia, ib, ja, jb,
                                          ka, kb, hx, ez_inc)
    hy, ihy = calculate_hy_field(ie, je, ke, hy, ihy, ex, ez,
                                 fj1, fi2, fk2, fi3, fk3)
    hy = calculate_hy_with_incident_field(ia, ib, ja, jb,
                                          ka, kb, hy, ez_inc)
    hz, ihz = calculate_hz_field(ie, je, ke, hz, ihz, ex, ey,
                                 fk1, fi2, fj2, fi3, fj3)

# Calculate the Fourier amplitude of the incident pulse
amp_in = np.sqrt(real_in ** 2 + imag_in ** 2)

# Calculate the Fourier amplitude of the total field
for m in range(number_of_frequencies):
    for j in range(ja, jb + 1):
        if gaz[ic, j, kc] < 1:
            amp[m, j] = 1 / (amp_in[m]) * sqrt(real_pt[m, ic, j, kc] ** 2 +
                                              imag_pt[m, ic, j, kc] ** 2)

# Plot Fig. 4.7
plt.rcParams['font.size'] = 12
plt.rcParams['grid.color'] = 'gray'
plt.rcParams['grid.linestyle'] = 'dotted'
fig = plt.figure(figsize=(8, 7))

X, Y = np.meshgrid(range(je), range(ie))

compare_array = np.arange(-8.5, 10.5, step=1)
x_array = np.arange(-20, 20, step=1)

# The data here was generated with the 3D Bessel function
  expansion program
compare_amp = np.array(
    [[0.074, 0.070, 0.064, 0.059, 0.054, 0.049, 0.044,
      0.038, 0.033, 0.028, 0.022, 0.017, 0.012, 0.007,
      0.005, 0.007, 0.012, 0.017, 0.022],
     [0.302, 0.303, 0.301, 0.294, 0.281, 0.263, 0.238,
      0.208, 0.173, 0.135, 0.095, 0.057, 0.036, 0.056,
      0.091, 0.126, 0.156, 0.182, 0.202],
```

```
    [0.329, 0.344, 0.353, 0.346, 0.336, 0.361, 0.436,
     0.526, 0.587, 0.589, 0.524, 0.407, 0.285, 0.244,
     0.300, 0.357, 0.360, 0.304, 0.208]])


def plot_amp(ax, data, compare, freq, scale):
    """Plot the Fourier transform amplitude at a specific frequency"""
    ax.plot(x_array, data, color='k', linewidth=1)
    ax.plot(compare_array, compare, 'ko', mfc='none', linewidth=1)
    plt.xlabel('cm')
    plt.ylabel('Amplitude')
    plt.xticks(np.arange(-5, 10, step=5))
    plt.xlim(-9, 9)
    plt.yticks(np.arange(0, 1, step=scale / 2))
    plt.ylim(0, scale)
    ax.text(20, 0.6, '{} MHz'.format(int(freq / 1e6)),
            horizontalalignment='center')


# Plot the results of the Fourier transform at each of the frequencies
scale = np.array((0.1, 0.5, 0.7))
for m in range(number_of_frequencies):
    ax = fig.add_subplot(3, 1, m + 1)
    plot_amp(ax, amp[m, :], compare_amp[m], freq[m], scale[m])

plt.tight_layout()
plt.show()
```

5

ADVANCED PYTHON FEATURES

This chapter explores some of the more advanced Python features. Previous programs were created to clearly illustrate the use of the derived equations within the text and, for the most part, be procedural for ease of understanding. This chapter also demonstrates alternatives to that code to take advantage of the fact that Python has many useful tools in the standard library. While by no means a complete tutorial, we will explore classes, overall structure, and interactive widgets to create good user-interfaces. These are just some of the features which make Python a useful programming language for simulations.

5.1 CLASSES

The class is a blueprint for creating objects and a fundamental building block in Python. A class is used to logically group data and functions that may operate on that data, which can also allow simplification of code and increase the readability. The functions that are part of a class are called methods. Here is an example. In three-dimensional simulation, a three-dimensional array was created for each E_x, E_y, and E_z. However, these are all components of the electric field and can logically be grouped together. Now we can create a class:

Electromagnetic Simulation Using the FDTD Method with Python, Third Edition.
Jennifer E. Houle and Dennis M. Sullivan.

Published 2020 by John Wiley & Sons, Inc.

```
class Field(object):
    """ This class creates a field in three directions.

    Attributes:
        x: Field strength in the x-direction
        y: Field strength in the y-direction
        z: Field strength in the z-direction
    """

    def __init__(self, x_cells, y_cells, z_cells, initial_value):
        """ Construct a Field object with an *x*, *y*, and
         *z* component
        Size of array generated is x_cells by y_cells by z_cells
        with given initial_value"""
        self.x=np.ones((x_cells, y_cells, z_cells)) * initial_value
        self.y=np.ones((x_cells, y_cells, z_cells)) * initial_value
        self.z=np.ones((x_cells, y_cells, z_cells)) * initial_value
```

This class definition does not create a field, but a blueprint we can later use to create as many independent fields as we would like. Upon creating an instance of `Field`, the `__init__` method is run immediately. This is where we construct the object and give it its initial data. Note that all methods in a class have a special first argument called `self`, which is a reference to the particular object whose method was called. Note that the parameter `self` is included as an input in the class definition, but it will not be given a value when the method is called (Python wires up "self" automatically). Next, we need to create an instance of the class, which is the object `e` (the electric field):

```
e = Field(IE, JE, KE, 0)
```

This sets up the three, three-dimensional electric field matrices and sets the initial values to zero. Now, the x, y, and z components can be accessed easily. For example, `e.x[0, 0, 0]` will access the first element of the electric field in the x-direction. Now we can pass the object `e` between functions, rather than having to manage `ex`, `ey`, and `ez` individually. This also improves readability, since all components are logically tied together.

In the program fd3d_4_3.py, there are quite a few fields, each in the x, y, and z directions. The code initializing the arrays is rather tedious and repetitive. A basic guideline of programming is "don't repeat yourself," so this should be addressed. Now we can create instances of the class `Field` instead and simplify the code.

```
e = Field(IE, JE, KE, 0)
i = Field(IE, JE, KE, 0)
d = Field(IE, JE, KE, 0)
id = Field(IE, JE, KE, 0)
h = Field(IE, JE, KE, 0)
```

```
    ih = Field(IE, JE, KE, 0)
    ga = Field(IE, JE, KE, 1)
    gb = Field(IE, JE, KE, 0)
```

This replaces the 24 lines of code in fd3d_4_3.py that initialize all the fields of interest.

Classes have another large advantage. Remember a class is used to group data and methods operating on that data. The above example was simply a grouping of the data. However, we may want to regularly manipulate that data. For example, if we are concerned with the total magnitude of the electric field and want to monitor it, we can add a method to our class.

```
def vector_magnitude(self):
    """ Return the vector magnitude of the field """
    return (self.x ** 2 + self.y ** 2 + self.z ** 2) ** (1 / 2)
```

Now to access the vector magnitude, all we need is to type:

```
e.vector_magnitude()
```

This is not particularly useful in fd3d_4_3.py, but demonstrates another feature of the class. To someone reading the code, what is happening is obvious when `e.vector_magnitude()` is called.

There are many additional class features and uses that are outside the scope of this discussion.

PROBLEM SET 5.1

1. Examine program fd3d_4_3.py to look for possible uses of other classes. Try creating the `Field` class and applying it to the program. Hint: Numba does not recognize this class, so it is important to pass each parameter into the function instead of the whole object in, that is,

   ```
   h.z, ih.z = calculate_hz_field(IE, JE, KE, h.z, ih.z, e.x, e.y)
   ```

   For standard functions (functions without the `@numba.jit` decorator), it is possible to simply pass in the object, that is `e`, and simply specify `e.x`, `e.y`, and `e.z` within the function. Note the `@jitclass` decorator may be added to the `Field` class, as shown in fd3d_5_1.py at the end of this chapter, to allow the flexibility of passing in the entire object.

5.1.1 Named Tuples

The custom class created in the previous section was used to group data and initialize arrays for a field. There are also many built-in classes in the Python standard library. Here we will discuss the "named tuple." The named tuple does not offer the versatility of a class created from scratch, but it is easy to use and sufficient for many purposes. Here we can use a named tuple to simplify our PML. First, we will create a new class:

```
PerfectlyMatchedLayer = namedtuple('PerfectlyMatchedLayer', (
    'fi1', 'fi2', 'fi3',
    'fj1', 'fj2', 'fj3',
    'fk1', 'fk2', 'fk3',
    'gi1', 'gi2', 'gi3',
    'gj1', 'gj2', 'gj3',
    'gk1', 'gk2', 'gk3',
))
```

Note that the "named tuple" function is called a "class factory." This factory function returns a class blueprint that can then be used to create objects. This function may sound complicated, but it is a useful shortcut to create simple blueprints without having to write a class definition manually. This again does not create a PML, but it specifies all our parameters.

Now we will create an instance of `PerfectlyMatchedLayer`, which will be called `pml`. Note the convention to capitalize the first letter of each word in the name of a class and to keep all letters lowercase in the instance of the class (the object).

```
    pml = PerfectlyMatchedLayer(
        fi1=np.zeros(IE), fi2=np.ones(IE), fi3=np.ones(IE),
        fj1=np.zeros(JE), fj2=np.ones(JE), fj3=np.ones(JE),
        fk1=np.zeros(KE), fk2=np.ones(KE), fk3=np.ones(KE),
        gi1=np.zeros(IE), gi2=np.ones(IE), gi3=np.ones(IE),
        gj1=np.zeros(JE), gj2=np.ones(JE), gj3=np.ones(JE),
        gk1=np.zeros(KE), gk2=np.ones(KE), gk3=np.ones(KE),
    )
```

This initializes the arrays for each PML parameter when the instance is created. In the case of `Field`, it was nice to have that initialization as part of the class, since it was called several times for each of our fields. The PML is only generated once, so it does not matter that we initialize the arrays on the line creating our instance.

Now the `calculate_pml_params` function in program fd3d_4_3.py can return a single `pml` object instead of 18 arrays. The only difference is that every time one of the PML arrays is used, it will need a `pml.` in front. For example,

```
        pml.fi1[n] = xn
        pml.fi1[IE - n - 1] = xn
```

While this does require extra characters, it improves readability, since it is now obvious when any of these parameters are used that they are part of the PML.

Also, the entire PML can be passed into and out of functions. Named tuples are recognized by Numba, so, for example, the function `calculate_dx_field` will now be

```
@numba.jit(nopython=True)
def calculate_dx_field(IE, JE, KE, dx, idx, hy, hz, pml):
    """ Calculate the Dx Field """
    for i in range(1, IE):
        for j in range(1, JE):
            for k in range(1, KE):
                curl_h = (hz[i, j, k] - hz[i, j - 1, k] - hy[i, j, k] +
                          hy[i, j, k - 1])
                idx[i, j, k] = idx[i, j, k] + curl_h
                dx[i, j, k] = pml.gj3[j] * pml.gk3[k] * dx[i, j, k] + \
                              pml.gj2[j] * pml.gk2[k] * \
                              (0.5 * curl_h + pml.gi1[i] * idx[i, j, k])
    return dx, idx
```

This is called by

```
    dx, idx = calculate_dx_field(IE, JE, KE, dx, idx, hy, hz, pml)
```

Obviously, there is still quite a bit of repetition in generating the PML. This will be addressed in the next section.

5.2 PROGRAM STRUCTURE

Ideally, code is broken into manageable pieces. Unlike a heavily imperative language like Fortran, higher-order languages like Python allow a mixture of functional and object-oriented styles to structure the code into small logical pieces, which can then be treated as black boxes and built into larger systems. This allows each function to be created, tested, and then used without needing to understand its details on each reading of the code.

5.2.1 Code Repetition

There are many reasons why reducing repetitive code can help. First, it is easier to check a single line of code. If a mistake is made or a change is needed, it is easier to change one line than to try to find every time a similar line of code was used. This is particularly important if a team is working on the code together. It is also easy, when done correctly, to use that bit of code again when it is part of a function. Finally, it can increase readability of the code.

We will begin to clean the PML code. In the program fd3d_4_3.py, the function `calculate_pml_params` was quite repetitive. By grouping those

TABLE 5.1 Breakdown of PML Parameters

PML Parameters	Array Size	Offset
*fi*1, *gi*2, *gi*3	ie	0
*fj*1, *gj*2, *gj*3	je	0
*fk*1, *gk*2, *gk*3	ke	0
*gi*1, *fi*2, *fi*3	ie	0.5
*gj*1, *fj*2, *fj*3	je	0.5
*gk*1, *fk*2, *fk*3	ke	0.5

parameters using a named tuple, we reduced the number of parameters being passed around and removed some clutter. Now we will simplify the function itself.

It is easy to see that much of the functionality is the same; *fi*1, *gi*2, and *gi*3 are calculated in the same manner as *fj*1, *gj*2, and *gj*3, as well as *fk*1, *gk*2, and *gk*3. The other nine parameters have similar parallels, except a half-cell offset is involved. We can break those parameters into "PML slices" as shown in Table 5.1.

Now we can create a function to calculate each PML slice.

```
def calculate_pml_slice(size, offset, pml_cells):
    """This initializes arrays and calculates a slice of the PML parameters
      (three of the parameters along one direction that use
        the same offset).
    fxl, gx2, gx3: offset = 0
    gxl, fx2, fx3: offset = 0.5 """

    distance = np.arange(pml_cells, 0, -1)
    xxn = (distance - offset) / pml_cells
    xn = 0.33 * (xxn ** 3)

    p1 = np.zeros(size)
    p2 = np.ones(size)
    p3 = np.ones(size)

    p1[:pml_cells] = xn
    p1[size - pml_cells: size] = np.flip(xn, 0)
    p2[:pml_cells] = 1 / (1 + xn)
    p2[size - pml_cells: size] = 1 / (1 + np.flip(xn, 0))
    p3[:pml_cells] = (1 - xn) / (1 + xn)
    p3[size - pml_cells: size] = \
        (1 - np.flip(xn, 0)) / (1 + np.flip(xn, 0))
return p1, p2, p3
```

Note the `for` loops were eliminated and instead built-in Numpy functions, such as `np.flip`, were used to make the calculations. To generate the entire set of PML parameters, only a few lines of code are needed:

```
fi1, gi2, gi3 = calculate_pml_slice(x_cells, 0, npml)
fj1, gj2, gj3 = calculate_pml_slice(y_cells, 0, npml)
fk1, gk2, gk3 = calculate_pml_slice(z_cells, 0, npml)
gi1, fi2, fi3 = calculate_pml_slice(x_cells, 0.5, npml)
gj1, fj2, fj3 = calculate_pml_slice(y_cells, 0.5, npml)
gk1, fk2, fk3 = calculate_pml_slice(z_cells, 0.5, npml)
```

These can then be used in the named tuple from the last section. Now, only 25 lines of code are used to generate the PML as opposed to 62 in the original program. These 25 lines of code may seem more complicated. However, it is easier to see how each of these PML parameters relate to the others. Most importantly, it is simple and quick to check the PML calculation. If there is an error or the calculation needs to be changed later, it is easier to make sure it is consistent for each of the PML parameters.

PROBLEM SET 5.2.1

1. Implement the more efficient PML code in program fd3d_4_3.py. Look for additional opportunities to remove redundancy and add clarification to the code.

5.2.2 Overall Structure

The overall structure of a program should be easy to follow. One way to increase readability is to create clear, explanatory variable and function names. Functions in a program should be named in a manner describing what they do without needing to dive into the code. Comments can support understanding, but the names alone should give an idea about a function's purpose. Similarly, the variable names should be clear and give a reason for existing. A terse variable or function name can be confusing when a few extra letters can make the code easy to follow.

Each function should be documented and easy to understand on its own. Try to avoid global variables; variables should be passed as arguments into each function to avoid mistakes by using the same variable names at different points in the program. This also makes it simple to create new programs using the same functions and makes the functions easier to understand because all inputs and outputs are clearly defined. In turn, these become easy-to-use black boxes.

The functions themselves should then be arranged to make a program understandable, allowing someone reading the code to quickly understand what the program does. Each individual function can then be examined for details of implementation.

The programs in this book are designed to be in a single file, since the programs are printed in the text. It is often a better practice to divide up the program into many files (modules) with similar functions placed together. These can then be accessed by other programs. However, due to the nature of this book, this will not be done.

Note that in the new version of the program, `__main__` will be used to call the function, `main`. This is a common Python idiom that ensures the script was directly executed (rather than imported) before running the code. It is not necessary here since we are not importing functions from other files, but it is a good habit nonetheless.

Since we have broken the code into smaller functions, we can display them in whatever order makes the most sense. The `main` program is called at the end of the program, so all functions are defined by the time we start executing code. Then, the main function can be placed at the top of the file, making it easy to see the basic building blocks of a program. Another useful feature is that variables of interest can be defined when the function is called to make them easy to adjust and see what should be modified. This keeps them separated and easy to locate, as is shown in program fd3d_5_1.py at the end of the chapter.

Returning to program fd3d_4_3.py, we will divide the program into the primary functions shown in Fig. 5.1. When looking at the function `main`, it is easy to follow what the program does at a high level. This breaks the code up into manageable pieces to write, test, organize, and read. Each of those main functions is similarly divided with each function easy to understand.

Program fd3d_5_1.py is a new version of fd3d_4_3.py with the new format. Many of the variable names have changed and additional comments have been added. However, the program produces the same results. Please note one additional function compared with fd3d_4_3.py, `save_outputs`, which saves the data that will be used in the following section.

PROBLEM SET 5.2.2

1. Run program fd3d_5_1.py and verify the results match fd3d_4_3.py.

5.3 INTERACTIVE WIDGETS

Many advantages of using a higher-order language like Python have already been shown. Python has a large ecosystem of open-source code including graphical plotting libraries, allowing a single program to compute and display data. These packages and libraries are constantly growing and expanding. Another feature of Python is the ability to create complete, user-friendly programs in a single language. The programs in this book require coding knowledge to run and modify. In some cases, the programmer may want to make a program anyone can use with minimal instruction, and Python can create a comprehensive program with interactivity. Interactive widgets include text

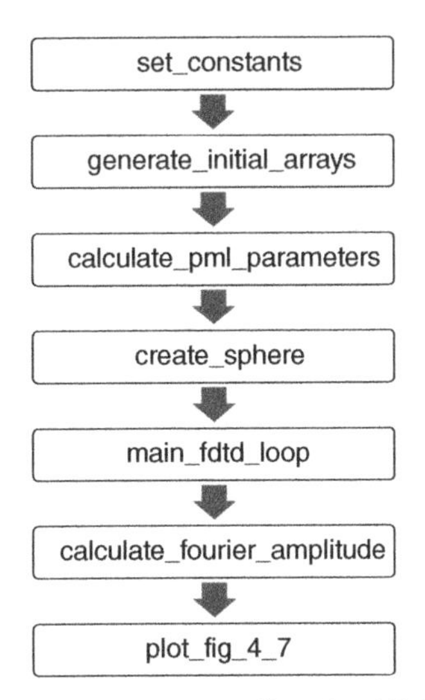

Figure 5.1 Main program flow for fd3d_5_1.py.

inputs, buttons, radio buttons, check buttons, and rectangle select, among others. This section gives basic information on matplotlib interactive widgets.

In this example, the outputs saved in the function `save_outputs` in fd3d_5_1.py are loaded, and the results at a single frequency are displayed. Radio buttons allow a user to select at which frequency the results are displayed.

The saved data are first loaded into named tuples, called `PlotParameters`.

```
PlotParameters = namedtuple('PlotParameters',
                            ['frequency','frequency_label',
                             'fdtd_location','fdtd_amplitude',
                             'bessel_location','bessel_amplitude'])
```

Each instance of `PlotParameters` groups the data for plotting at a single frequency. It includes a `frequency_label` that is used to make the frequency data display nicely on the radio buttons. Each of these instances will be stored in a list that collects the `PlotParameters` across all frequencies. This list can easily be passed between functions and the data will be kept together for a given frequency. The list is generated using a list comprehension as follows, using the saved output data from fd3d_5_1.py:

```
PlotParameters(
    frequency=frequency,
```

```
        frequency_label="{} MHz".format(int(frequency / 1e6)),
        fdtd_location=fdtd_x_axis,
        fdtd_amplitude=fdtd_amp[i, :],
        bessel_location=bessel_x_axis,
        bessel_amplitude=bessel_amp[i, :]
    )
    for i, frequency in enumerate(frequencies)
```

We will create a `Controller` class to manage the interactivity (e.g., this `Controller` will keep track of what frequency the user has selected). An instance of this class is created and includes a `RadioButtons` widget that allows a user to interact with the program. A `Controller` class is not the only way to approach this problem, but the `Controller` class makes it easier to share data between methods without needing to explicitly pass every shared variable across all functions. The class created for `Controller` has a similar structure to the class created in Section 5.1.

```
class Controller:
    """ This class creates the controller for the interactive widgets.
    """

    def __init__(self, figure, current_ax, plot_parameters, fdtd_plot,
                 bessel_plot):
        self.figure = figure
        self.current_ax = current_ax
        self.plot_parameters = plot_parameters
        self.fdtd_plot = fdtd_plot
        self.bessel_plot = bessel_plot

        self.selected_freq_label = \
            plot_parameters[0].frequency_label # Initial frequency

        # This sets up the radio buttons
        radio_axes = plt.axes([0.03, 0.55, 0.15, 0.20]) # Radio button axis
        self.radio = RadioButtons(
            ax=radio_axes,
            labels=[output.frequency_label for output in plot_parameters]
        )
        self.radio.on_clicked(self.on_radio_select)

    def on_radio_select(self, label_selected):
        """On radio select, the new frequency is selected
        and plots are redrawn"""
        self.selected_freq_label = label_selected
        self.redraw()

    def redraw(self):
        """Redraw the figure based on current_freq"""
        for plot_parameter in self.plot_parameters:
```

```
            if self.selected_freq_label == plot_parameter.
              frequency_label:
              self.fdtd_plot.set_data(plot_parameter.fdtd_location,
                                      plot_parameter. fdtd_amplitude)
              self.bessel_plot.set_data(plot_parameter. bessel_
                                        location,
                                      plot_parameter. bessel_amplitude)
        self.current_ax.relim()
        self.current_ax.autoscale_view(True, True, True)
        plt.draw()
```

Notice there are three methods that are part of Controller: `__init__`, `on_radio_select`, and `redraw`.

As with `Field` in Section 5.1, the `__init__` method shows what happens when the instance of the class is created. Any parameters that were passed in are assigned to that instance using `self`. The first two parameters, `figure` and `current_ax`, keep track of the final image and the axes, respectively. Note this refers to "axes," which is the information about a given plot, and not axis (as in the x or y axis); this is a matplotlib naming convention. These parameters can be saved when a plot is generated, which is done in fd3d_5_2.py at the end of this chapter.

Recall that `plot_parameters` is a list of instances of the class `PlotParameters`. We store this list as an attribute in the class so it can be easily accessed by the methods in `Controller`. The next two attributes, `fdtd_plot` and `bessel_plot`, are the two plots displayed (FDTD results and Bessel results, respectively). Then, the selected frequency (`selected_freq_label`) is initialized to the `frequency` value from the first instance of `PlotParameters`.

The radio buttons are created after the axes object, which places the radio buttons within the window:

```
    radio_axes = plt.axes([0.03, 0.55, 0.15, 0.20]).
```

Next, the radio buttons are put on the axes that was just created by creating an instance of the class `RadioButtons` called `radio`. This is accomplished with the line:

```
        self.radio = RadioButtons(
            ax=radio_axes,
            labels=[output.frequency_label for output in
              plot_parameters]
        )
```

The list called `labels` (a part of the `RadioButtons` class) stores each `frequency_label` that is part of `plot_parameters`. These `labels` set up the string that is displayed by each radio button, one for each frequency. Finally, we set the radio button's `on_clicked` parameter:

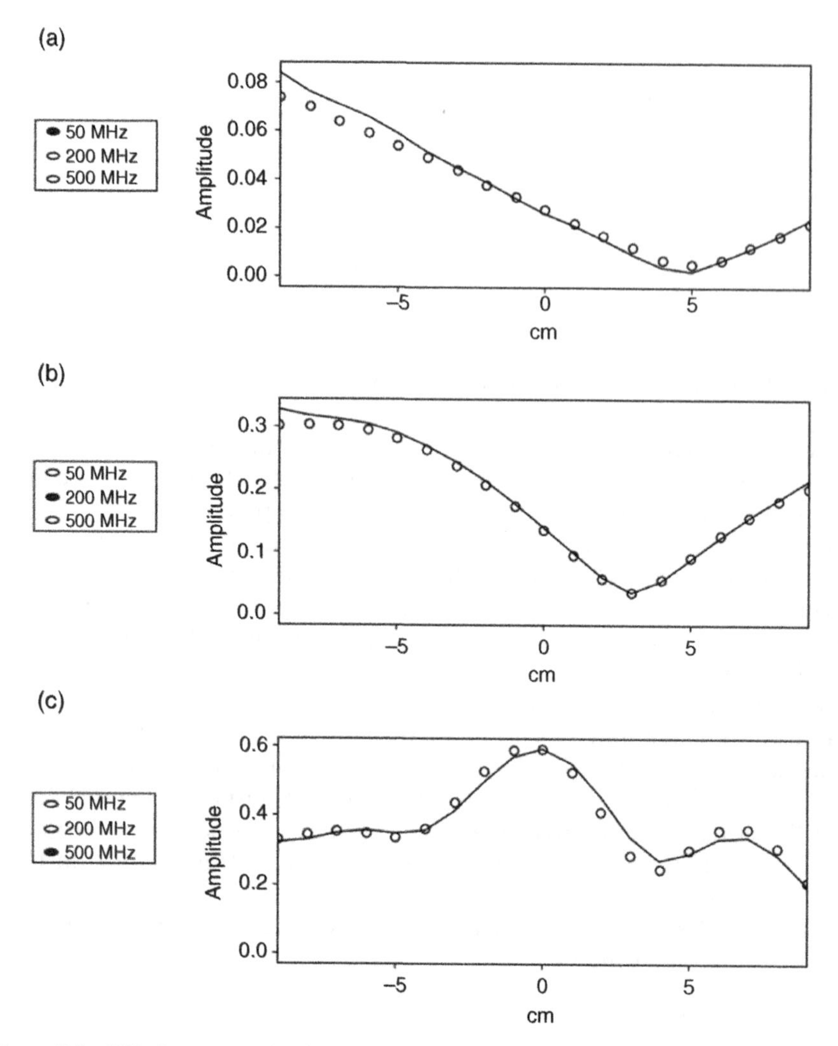

Figure 5.2 Window created using the `Controller` class. This displays the output data from fd3d_5_1.py for one frequency at a time using the interactive radio buttons on the left. Results are displayed at (a) 50, (b) 200, and (c) 500 MHz.

```
self.radio.on_clicked(self.on_radio_select)
```

to control what happens when the radio button is clicked. Clicking a radio button will call the method we passed to `on_clicked` (in this case the `on_radio_select` method we created) when one of the radio buttons is clicked by a user. It will also pass in the button label (from the list `labels`) associated with the clicked radio button.

The method `on_radio_select` will select a new frequency and update the plot. It has two inputs: itself (`self`) and `label_selected`. The input `label_selected` was passed in from the `on_clicked` method previously described. The method `on_radio_select` then calls the final method that is part of `Controller`, `redraw`. The `redraw` method does exactly what its name indicates: redraw the figure with the newly selected data. Figure 5.2 shows what the window created using `Controller` looks like.

This very simple example demonstrates the potential of interactive widgets. Another example will be given in the next chapter for a practical application when the user does not necessarily understand what the program does or how the code works, but an easy-to-use interface allows the user to explore and manipulate data. There are a number of ways to package Python code in an executable application that may be installed and launched. Then a user does not even need to install Python to use the code.

PROBLEM SET 5.3

1. The code for creating the interactive program of Fig. 5.2 is in fd3d_5_2.py. Create a button that will display only the FDTD data, but not the Bessel comparison data (which was displayed as circles). Remember the input data for fd3d_5_2.py is generated in fd3d_5_1.py.

```
""" fd3d_5_1.py: 3d FDTD
Chapter 4 Section 3
3D FDTD simulation of a plane wave on a dielectric sphere
Refactored
"""

from collections import namedtuple
from math import exp, sqrt, cos, sin

import numba
import numpy as np
from matplotlib import pyplot as plt


def main(nsteps, num_freq, freq, dims):
    c = set_constants(freq)

    real_in, imag_in, real_pt, imag_pt, amp, e, i, d, id, h, ih, \
    hx_inc, ez_inc = generate_initial_arrays(dims, num_freq)

    pml = calculate_pml_params(dims, npml=8)
    ga, gb = create_sphere(c, dims)
```

```
    real_in, imag_in, num_freq, real_pt, imag_pt = \
        main_fdtd_loop(nsteps, c, dims, ez_inc, num_freq,
                       real_in, imag_in,
                       d, h, pml, e, i, ih, id, ga, gb, hx_inc, real_pt,
                       imag_pt)

    amp = calculate_fourier_amplitude(dims, real_in, imag_in, ga,
                                      num_freq, amp, real_pt,
                                      imag_pt)
    plot_fig_4_7(amp, num_freq, freq)


# the @numba.jitclass decorator allows us to pass our custom
# class into a `numba.jit`ed function
@numba.jitclass([
    ('x', numba.float32[:, :, :]),
    ('y', numba.float32[:, :, :]),
    ('z', numba.float32[:, :, :])
])
class Field:
    """ This class creates a field in three directions.

    Attributes:
        x: Field strength in the x-direction
        y: Field strength in the y-direction
        z: Field strength in the z-direction
    """

    def __init__(self, x_cells, y_cells, z_cells, initial_value):
        """ Return a Field object with an *x*, *y*, and
          *z* component"""
        self.x = np.ones((x_cells, y_cells, z_cells),
                         dtype=np.float32) * initial_value
        self.y = np.ones((x_cells, y_cells, z_cells),
                         dtype=np.float32) * initial_value
        self.z = np.ones((x_cells, y_cells, z_cells),
                         dtype=np.float32) * initial_value


@numba.jitclass([
    ('x', numba.int16),
    ('y', numba.int16),
    ('z', numba.int16),
    ('x_center', numba.int16),
    ('y_center', numba.int16),
    ('z_center', numba.int16),
    ('xa', numba.int16),
```

```
    ('ya', numba.int16),
    ('za', numba.int16),
    ('xb', numba.int16),
    ('yb', numba.int16),
    ('zb', numba.int16)
])
class Dimensions:
    """ This class keeps track of the problem space in
      three directions.

    Attributes:
        x: Number of cells in the x-direction
        y: Number of cells in the y-direction
        z: Number of cells in the z-direction
        x_center: Index for the center cell in the x-direction
        y_center: Index for the center cell in the y-direction
        z_center: Index for the center cell in the z-direction
        xa: Scattered/total field lower boundary in the x-direction
        ya: Scattered/total field lower boundary in the y-direction
        za: Scattered/total field lower boundary in the z-direction
        xb: Scattered/total field upper boundary in the x-direction
        yb: Scattered/total field upper boundary in the y-direction
        zb: Scattered/total field upper boundary in the z-direction
    """

    def__init__(self, x, y, z, xa, ya, za):
       self.x = x
       self.y = y
       self.z = z

       self.x_center = int(self.x / 2)
       self.y_center = int(self.y / 2)
       self.z_center = int(self.z / 2)

       self.xa = xa
       self.ya = ya
       self.za = za

       self.xb = self.x - self.xa - 1
       self.yb = self.y - self.ya - 1
       self.zb = self.z - self.za - 1


Constants = namedtuple('Constants', (
    'ddx', 'dt', 'arg',
    't0', 'spread'
))
```

```
PerfectlyMatchedLayer = namedtuple('PerfectlyMatchedLayer', (
    'fi1', 'fi2', 'fi3',
    'fj1', 'fj2', 'fj3',
    'fk1', 'fk2', 'fk3',
    'gi1', 'gi2', 'gi3',
    'gj1', 'gj2', 'gj3',
    'gk1', 'gk2', 'gk3',
))


def set_constants(freq):
    """ Set up constants that do not change for the
      entire simulation"""

    ddx = 0.01  # Cell size
    dt = ddx / 6e8  # Time step size
    arg = 2 * np.pi * freq * dt

    # Pulse Parameters
    t0 = 20
    spread = 8

    c = Constants(
        ddx, dt, arg,
        t0, spread
    )
    return c


def generate_initial_arrays(dims, num_freq):
    """Generate the arrays that will be used in the program"""

    real_in = np.zeros(num_freq)
    imag_in = np.zeros(num_freq)
    real_pt = np.zeros((num_freq, dims.x, dims.y, dims.z))
    imag_pt = np.zeros((num_freq, dims.x, dims.y, dims.z))
    amp = np.zeros((num_freq, dims.y))

    # Generate the initial field objects
    e = Field(dims.x, dims.y, dims.z, 0)
    i = Field(dims.x, dims.y, dims.z, 0)
    d = Field(dims.x, dims.y, dims.z, 0)
    id = Field(dims.x, dims.y, dims.z, 0)
    h = Field(dims.x, dims.y, dims.z, 0)
    ih = Field(dims.x, dims.y, dims.z, 0)
```

```
    hx_inc = np.zeros(dims.y)
    ez_inc = np.zeros(dims.y)

    return real_in, imag_in, real_pt, imag_pt, amp, e, i,
      d, id, h, ih, hx_inc, ez_inc


def calculate_pml_params(dims, npml):
    """ Creates the Perfectly Matched Layer object """

    fi1, gi2, gi3 = calculate_pml_slice(dims.x, 0, npml)
    fj1, gj2, gj3 = calculate_pml_slice(dims.y, 0, npml)
    fk1, gk2, gk3 = calculate_pml_slice(dims.z, 0, npml)

    gi1, fi2, fi3 = calculate_pml_slice(dims.x, 0.5, npml)
    gj1, fj2, fj3 = calculate_pml_slice(dims.y, 0.5, npml)
    gk1, fk2, fk3 = calculate_pml_slice(dims.z, 0.5, npml)

    pml = PerfectlyMatchedLayer(
        fi1=fi1, fi2=fi2, fi3=fi3,
        fj1=fj1, fj2=fj2, fj3=fj3,
        fk1=fk1, fk2=fk2, fk3=fk3,

        gi1=gi1, gi2=gi2, gi3=gi3,
        gj1=gj1, gj2=gj2, gj3=gj3,
        gk1=gk1, gk2=gk2, gk3=gk3,
    )

    return pml


def calculate_pml_slice(size, offset, pml_cells):
    """ This initializes arrays and calculates a slice of
      the PML parameters
    (three of the parameters along one direction that use
      the same offset).
    fx1, gx2, gx3: offset = 0
    gx1, fx2, fx3: offset = 0.5 """

    distance = np.arange(pml_cells, 0, -1)
    xxn = (distance - offset) / pml_cells
    xn = 0.33 * (xxn ** 3)

    p1 = np.zeros(size)
    p2 = np.ones(size)
    p3 = np.ones(size)
```

```
    p1[:pml_cells] = xn
    p1[size - pml_cells: size] = np.flip(xn, 0)
    p2[:pml_cells] = 1 / (1 + xn)
    p2[size - pml_cells: size] = 1 / (1 + np.flip(xn, 0))
    p3[:pml_cells] = (1 - xn) / (1 + xn)
    p3[size - pml_cells: size] = \
        (1 - np.flip(xn, 0)) / (1 + np.flip(xn, 0))

    return p1, p2, p3


def create_sphere(c, dims):
    """ Specify the parameters of the sphere, then
      generate the sphere """

    epsz = 8.854e-12

    # Specify sphere parameters
    epsilon = np.ones(2)
    sigma = np.zeros(2)
    epsilon[1] = 30
    sigma[1] = 0.3
    radius = 10

    # Generate Field objects
    ga = Field(dims.x, dims.y, dims.z, 1)
    gb = Field(dims.x, dims.y, dims.z, 0)

    # Generate the sphere
    ga.x, gb.x = create_sphere_single_direction(c, dims,
      0.5, 0, 0, epsilon, sigma, radius, ga.x, gb.x, epsz)
    ga.y, gb.y = create_sphere_single_direction(c, dims,
      0, 0.5, 0, epsilon, sigma, radius, ga.y, gb.y, epsz)
    ga.z, gb.z = create_sphere_single_direction(c, dims,
      0, 0, 0.5, epsilon, sigma, radius, ga.z, gb.z, epsz)
    return ga, gb


def create_sphere_single_direction(c, dims, offset_x,
  offset_y, offset_z, epsilon, sigma, radius, ga, gb, epsz):
    """ Create the sphere in one direction (i.e. x, y, or
      z direction) """

    for x in range(dims.xa, dims.xb + 1):
        for y in range(dims.ya, dims.yb + 1):
            for z in range(dims.za, dims.zb + 1):
                eps = epsilon[0]
```

```
                cond = sigma[0]
                xdist = dims.x_center - x - offset_x
                ydist = dims.y_center - y - offset_y
                zdist = dims.z_center - z - offset_z
                dist = sqrt(xdist ** 2 + ydist ** 2 + zdist ** 2)
                if dist <= radius:
                    eps = epsilon[1]
                    cond = sigma[1]
                ga[x, y, z] = 1 / (eps + (cond * c.dt / epsz))
                gb[x, y, z] = cond * c.dt / epsz

    return ga, gb


def main_fdtd_loop(nsteps, c, dims, ez_inc, num_freq,
  real_in, imag_in, d, h, pml, e, i, ih, id, ga, gb, hx_inc,
      real_pt, imag_pt):
    # Absorbing Boundary Conditions for the Plane Wave
    boundary_low = [0, 0]
    boundary_high = [0, 0]

    for time_step in range(1, nsteps + 1):
        ez_inc = calculate_incident_buffer(ez_inc,
          hx_inc, dims.y)

        real_in, imag_in = \
            fourier_transform_inc_field(real_in, imag_in, num_freq,
                                        time_step, dims.ya, ez_inc,
                                        c.arg)

        ez_inc, boundary_high, boundary_low = \
            absorbing_bound_cond(ez_inc, boundary_high,
                                 boundary_low, dims.y)

        d.x, id.x = calculate_dx_field(dims, d.x, id.x, h, pml)
        d.y, id.y = calculate_dy_field(dims, d.y, id.y, h, pml)
        d.z, id.z = calculate_dz_field(dims, d.z, id.z, h, pml)

        ez_inc[3] = add_source_in_gap(time_step, c.t0, c.
          spread)

        d.y = calculate_inc_dy_field(dims, d.y, hx_inc)
        d.z = calculate_inc_dz_field(dims, d.z, hx_inc)

        e, i = calculate_e_fields(dims, d, ga, gb, e, i)
```

```
        real_pt, imag_pt \
            = calculate_fourier_transform_ex(c, dims,
              num_freq, real_pt, imag_pt, e.z, time_step)

        hx_inc = calculate_hx_inc(dims.y, hx_inc, ez_inc)
        h.x, ih.x = calculate_hx_field(dims, h.x, ih.x, e, pml)
        h.x = calculate_hx_with_incident_field(dims, h.x, ez_inc)
        h.y, ih.y = calculate_hy_field(dims, h.y, ih.y, e, pml)
        h.y = calculate_hy_with_incident_field(dims, h.y, ez_inc)
        h.z, ih.z = calculate_hz_field(dims, h.z, ih.z, e, pml)

    return real_in, imag_in, num_freq, real_pt, imag_pt


def calculate_incident_buffer(ez_inc, hx_inc, dims_y):
    """ Calculate Ez using the incident fields """
    for y in range(1, dims_y - 1):
        ez_inc[y] = ez_inc[y] + 0.5 * (hx_inc[y - 1] - hx_inc[y])

    return ez_inc


def fourier_transform_inc_field(real_in, imag_in,
  num_freq, time_step, dims_y, ez_inc, arg):
    """ Fourier transform of the incident field """
    for m in range(num_freq):
        real_in[m] = real_in[m] + cos(arg[m] * time_step) \
                     * ez_inc[dims_y - 1]
        imag_in[m] = imag_in[m] - sin(arg[m] * time_step) \
                     * ez_inc[dims_y - 1]

    return real_in, imag_in


def absorbing_bound_cond(ez_inc, boundary_high, boundary_low,
  dims_y):
    """Absorbing Boundary Conditions for the incident array"""
    ez_inc[0] = boundary_low.pop(0)
    boundary_low.append(ez_inc[1])

    ez_inc[dims_y - 1] = boundary_high.pop(0)
    boundary_high.append(ez_inc[dims_y - 2])

    return ez_inc, boundary_high, boundary_low
```

```
@numba.jit(nopython=True)
def calculate_dx_field(dims, dx, idx, h, pml):
    """ Calculate the Dx Field.
    Implementing equation analogous to Eq. (4.6) """
    for x in range(1, dims.x):
        for y in range(1, dims.y):
            for z in range(1, dims.z):
                curl_h = (h.z[x, y, z] - h.z[x, y - 1, z] -
                          h.y[x, y, z] + h.y[x, y, z - 1])
                idx[x, y, z] = idx[x, y, z] + curl_h
                dx[x, y, z] = pml.gj3[y] * pml.gk3[z] * dx
                  [x, y, z] + \
                              pml.gj2[y] * pml.gk2[z] * \
                              (0.5 * curl_h + pml.gi1[x] *
                                idx[x, y, z])
    return dx, idx


@numba.jit(nopython=True)
def calculate_dy_field(dims, dy, idy, h, pml):
    """ Calculate the Dy Field.
    Implementing equation analogous to Eq. (4.6) """
    for x in range(1, dims.x):
        for y in range(1, dims.y):
            for z in range(1, dims.z):
                curl_h = (h.x[x, y, z] - h.x[x, y, z - 1] -
                          h.z[x, y, z] + h.z[x - 1, y, z])
                idy[x, y, z] = idy[x, y, z] + curl_h
                dy[x, y, z] = pml.gi3[x] * pml.gk3[z] * dy
                  [x, y, z] + \
                              pml.gi2[x] * pml.gk2[z] * \
                              (0.5 * curl_h + pml.gj1[y] *
                                idy[x, y, z])
    return dy, idy


@numba.jit(nopython=True)
def calculate_dz_field(dims, dz, idz, h, pml):
    """ Calculate the Dz Field.
    Implementing Eq. (4.6) """
    for x in range(1, dims.x):
        for y in range(1, dims.y):
            for z in range(1, dims.z):
                curl_h = (h.y[x, y, z] - h.y[x - 1, y, z] -
                          h.x[x, y, z] + h.x[x, y - 1, z])
                idz[x, y, z] = idz[x, y, z] + curl_h
```

```
                dz[x, y, z] = pml.gi3[x] * pml.gj3[y] * dz
                  [x, y, z] + \
                              pml.gi2[x] * pml.gj2[y] * \
                              (0.5 * curl_h + pml.gk1[z] *
                                idz[x, y, z])
    return dz, idz


def add_source_in_gap(time_step, t0, spread):
    """Add the source at the gap to generate a pulse"""
    pulse = exp(-0.5 * ((t0 - time_step) / spread) ** 2)
    return pulse


@numba.jit(nopython=True)
def calculate_inc_dy_field(dims, dy, hx_inc):
    """ Calculate the incident Dy Field
    Implementing Eq. (4.7a) and (4.7b) """
    for x in range(dims.xa, dims.xb + 1):
        for y in range(dims.ya, dims.yb + 1):
            dy[x, y, dims.za] = dy[x, y, dims.za] - 0.5 *
              hx_inc[y]
            dy[x, y, dims.zb + 1] = dy[x, y, dims.zb + 1] +
              0.5 * hx_inc[y]
    return dy


@numba.jit(nopython=True)
def calculate_inc_dz_field(dims, dz, hx_inc):
    """ Calculate the incident Dz Field
    Implementing equations analogous to Eq. (3.26a) and
      (3.26b) """
    for x in range(dims.xa, dims.xb + 1):
        for z in range(dims.za, dims.zb + 1):
            dz[x, dims.ya, z] = dz[x, dims.ya, z] + 0.5 *
              hx_inc[dims.ya - 1]
            dz[x, dims.yb, z] = dz[x, dims.yb, z] - 0.5 *
              hx_inc[dims.yb]
    return dz


@numba.jit(nopython=True)
def calculate_e_fields(dims, d, ga, gb, e, i):
    """ Calculate the E field from the D field
   Implementing equations analogous to Eq. (2.9b) and (2.9c) """
    for x in range(0, dims.x):
        for y in range(0, dims.y):
```

```
            for z in range(0, dims.z):
                e.x[x, y, z] = ga.x[x, y, z] * (d.x[x, y,
                  z] - i.x[x, y, z])
                i.x[x, y, z] = i.x[x, y, z] + gb.x[x, y, z]
                  * e.x[x, y, z]
                e.y[x, y, z] = ga.y[x, y, z] * (d.y[x, y,
                  z] - i.y[x, y, z])
                i.y[x, y, z] = i.y[x, y, z] + gb.y[x, y, z]
                  * e.y[x, y, z]
                e.z[x, y, z] = ga.z[x, y, z] * (d.z[x, y,
                  z] - i.z[x, y, z])
                i.z[x, y, z] = i.z[x, y, z] + gb.z[x, y, z]
                  * e.z[x, y, z]
    return e, i


@numba.jit(nopython=True)
def calculate_fourier_transform_ex(c, dims, number_of_frequencies,
                                   real_pt, imag_pt, ez, time_step):
    """ Calculate the Fourier transform of Ez"""
    for x in range(0, dims.x):
        for y in range(0, dims.y):
            for m in range(0, number_of_frequencies):
                real_pt[m, x, y] = \
                 real_pt[m, x, y] + cos(c.arg[m] * time_step) * \
                    ez[x, y, dims.z_center]
                imag_pt[m, x, y] = \
                    imag_pt[m, x, y] - sin(c.arg[m] * time_step) * \
                    ez[x, y, dims.z_center]
    return real_pt, imag_pt


@numba.jit(nopython=True)
def calculate_hx_field(dims, hx, ihx, e, pml):
    """ Calculate the Hx field
    Implementing equations analogous to Eq. (3.24a)-(3.24c)
    and described in Section 4.2"""
    for x in range(0, dims.x):
        for y in range(0, dims.y - 1):
            for z in range(0, dims.z - 1):
                curl_e = (e.y[x, y, z + 1] - e.y[x, y, z] -
                          e.z[x, y + 1, z] + e.z[x, y, z])
                ihx[x, y, z] = ihx[x, y, z] + curl_e
                hx[x, y, z] = pml.fj3[y] * pml.fk3[z] * hx
                  [x, y, z] + \
                            pml.fj2[y] * pml.fk2[z] * 0.5 * \
                          (curl_e + pml.fi1[x] * ihx[x, y, z])
    return hx, ihx
```

```
@numba.jit(nopython=True)
def calculate_hy_field(dims, hy, ihy, e, pml):
    """ Calculate the Hy field
    Implementing equations analogous to Eq. (3.24a)-(3.24c)
    and described in Section 4.2"""
    for x in range(0, dims.x - 1):
        for y in range(0, dims.y):
            for z in range(0, dims.z - 1):
                curl_e = (e.z[x + 1, y, z] - e.z[x, y, z] -
                          e.x[x, y, z + 1] + e.x[x, y, z])
                ihy[x, y, z] = ihy[x, y, z] + curl_e
                hy[x, y, z] = pml.fi3[x] * pml.fk3[z] * hy[x, y, z] + \
                              pml.fi2[x] * pml.fk2[z] * 0.5 * \
                              (curl_e + pml.fj1[y] * ihy[x, y, z])
    return hy, ihy


@numba.jit(nopython=True)
def calculate_hz_field(dims, hz, ihz, e, pml):
    """ Calculate the Hz field
    Implementing equations analogous to Eq. (3.24a)-(3.24c)
    and described in Section 4.2"""
    for x in range(0, dims.x - 1):
        for y in range(0, dims.y - 1):
            for z in range(0, dims.z):
                curl_e = (e.x[x, y + 1, z] - e.x[x, y, z] -
                          e.y[x + 1, y, z] + e.y[x, y, z])
                ihz[x, y, z] = ihz[x, y, z] + curl_e
                hz[x, y, z] = pml.fi3[x] * pml.fj3[y] * hz[x, y, z] + \
                              pml.fi2[x] * pml.fj2[y] * 0.5 * \
                              (curl_e + pml.fk1[z] * ihz[x, y, z])
    return hz, ihz


@numba.jit(nopython=True)
def calculate_hx_inc(dims_y, hx_inc, ez_inc):
    """ Calculate incident Hx field"""
    for y in range(0, dims_y - 1):
        hx_inc[y] = hx_inc[y] + 0.5 * (ez_inc[y] - ez_inc[y + 1])

    return hx_inc
```

```
@numba.jit(nopython=True)
def calculate_hx_with_incident_field(dims, hx, ez_inc):
    """ Calculate Hx with incident Ez
    Implementing equations analogous to Eq. (3.27a) and (3.27b) """
    for x in range(dims.xa, dims.xb + 1):
        for z in range(dims.za, dims.zb + 1):
            hx[x, dims.ya - 1, z] = hx[x, dims.ya - 1, z] + \
                                    0.5 * ez_inc[dims.ya]
            hx[x, dims.yb, z] = hx[x, dims.yb, z] - \
                                0.5 * ez_inc[dims.yb]

    return hx


@numba.jit(nopython=True)
def calculate_hy_with_incident_field(dims, hy, ez_inc):
    """ Calculate Hy with incident Ez
    Implementing equations analogous to Eq. (3.28a) and (3.28b) """
    for y in range(dims.ya, dims.yb + 1):
        for z in range(dims.za, dims.zb + 1):
           hy[dims.xa - 1, y, z] = hy[dims.xa - 1, y, z] - 0.5 *
             ez_inc[y]
           hy[dims.xb, y, z] = hy[dims.xb, y, z] + 0.5 * ez_inc[y]

    return hy


def calculate_fourier_amplitude(dims, real_in, imag_in, ga,
  num_freq, amp,
                                real_pt, imag_pt):
    """ Calculate the Fourier amplitude of the total field"""

    # Calculate the Fourier amplitude of the incident pulse
    amp_in = np.sqrt(real_in ** 2 + imag_in ** 2)

    # Calculate the Fourier amplitude of the total field
    for m in range(num_freq):
        for y in range(dims.ya, dims.yb + 1):
            if ga.z[dims.x_center, y, dims.z_center] < 1:
                amp[m, y] = \
                    1 / (amp_in[m]) * \
                    sqrt(real_pt[m, dims.x_center, y, dims.
                      z_center] ** 2 +
                         imag_pt[m, dims.x_center, y, dims.
                           z_center] ** 2)
    return amp
```

```
def plot_fig_4_7(amp, num_freq, freq):
    """Plot Fig. 4.7"""

    plt.rcParams['font.size'] = 12
    plt.rcParams['grid.color'] = 'gray'
    plt.rcParams['grid.linestyle'] = 'dotted'
    fig = plt.figure(figsize=(8, 7))

    compare_array = np.arange(-9, 10, step=1)
    x_array = np.arange(-20, 20, step=1)

    # The data here was generated with the 3D Bessel function
    # expansion program
    compare_amp = np.array(
        [[0.074, 0.070, 0.064, 0.059, 0.054, 0.049, 0.044,
          0.038, 0.033, 0.028, 0.022, 0.017, 0.012, 0.007,
          0.005, 0.007, 0.012, 0.017, 0.022],
         [0.302, 0.303, 0.301, 0.294, 0.281, 0.263, 0.238,
          0.208, 0.173, 0.135, 0.095, 0.057, 0.036, 0.056,
          0.091, 0.126, 0.156, 0.182, 0.202],
         [0.329, 0.344, 0.353, 0.346, 0.336, 0.361, 0.436,
          0.526, 0.587, 0.589, 0.524, 0.407, 0.285, 0.244,
          0.300, 0.357, 0.360, 0.304, 0.208]])

    # Plot the results of the Fourier transform at each of the
      frequencies
    scale = np.array((0.1, 0.5, 0.7))
    for m in range(num_freq):
        ax = fig.add_subplot(3, 1, m + 1)
        plot_amp(ax, amp[m, :], compare_amp[m], freq[m], scale
          [m], x_array,
                 compare_array)

    save_outputs(freq, amp, compare_amp, x_array, compare_array)

    plt.tight_layout()
    plt.show()

    return


def plot_amp(ax, data, compare, freq, scale, x_array, compare_array):
    """Plot the Fourier transform amplitude at a specific frequency"""
    ax.plot(x_array, data, color='k', linewidth=1)
    ax.plot(compare_array, compare, 'ko', mfc='none', linewidth=1)
    plt.xlabel('cm')
    plt.ylabel('Amplitude')
    plt.xticks(np.arange(-5, 10, step=5))
```

```
    plt.xlim(-9, 9)
    plt.yticks(np.arange(0, 1, step=scale / 2))
    plt.ylim(0, scale)
    ax.text(20, 0.6, '{} MHz'.format(int(freq / 1e6)),
            horizontalalignment='center')


def save_outputs(freq, amp, compare_amp, x_array,
compare_array):
    """ Save the numpy arrays of amplitude and
    the Bessel result's amplitude """
    np.save('fdtd_amp', amp)
    np.save('bessel_amp', compare_amp)
    np.save('fdtd_x_axis', x_array)
    np.save('bessel_x_axis', compare_array)
    np.save('frequencies', freq)
    return


if __name__ == '__main__':
    dims = Dimensions(
        x=40, y=40, z=40,
        xa=7, ya=7, za=7
    )

    main(
        nsteps=500,
        num_freq=3,
        freq=np.array((50e6, 200e6, 500e6)),
        dims=dims
    )


""" fd3d_5_2.py: Interactive

Use interactive widgets to view output data from fdtd_5_1
"""
from collections import namedtuple

import matplotlib.pyplot as plt
import numpy as np
from matplotlib.widgets import RadioButtons

plt.rcParams['toolbar'] = 'None'


def main():
    plot_parameters = load_plot_parameters()
```

```
    figure, current_ax, fdtd_plot, bessel_plot = plot_fig_4_7(
        plot_parameters[0])
    plt.subplots_adjust(left=0.33)

    # Create an instance of the class 'Controller'
    radio_controller = Controller(figure, current_ax,
      plot_parameters, fdtd_plot, bessel_plot)
    plt.show()


class Controller:
    """This class creates the controller for the interactive widgets.

    Attributes:
        figure: figure
        current_ax: axes for the figure
        plot_parameters: data saved from fd3d_5_1 (frequency, fdtd
                    amplitude, Bessel amplitude, and x-axis data)
        fdtd_plot = fdtd data plot
        bessel_plot = Bessel data plot
    """

  def __init__(self, figure, current_ax, plot_parameters, fdtd_plot,
             bessel_plot):
        self.figure = figure
        self.current_ax = current_ax
        self.plot_parameters = plot_parameters
        self.fdtd_plot = fdtd_plot
        self.bessel_plot = bessel_plot

        self.selected_freq_label = \
           plot_parameters[0].frequency_label # Initial frequency

        # This sets up the radio buttons
        radio_axes = plt.axes([0.03, 0.55, 0.15, 0.20])
        # Radio button axis
        self.radio = RadioButtons(
            ax=radio_axes,
            labels=[output.frequency_label for output in
              plot_parameters]
        )
        self.radio.on_clicked(self.on_radio_select)

    def on_radio_select(self, label_selected):
        """On radio select, the new frequency is selected
        and plots are redrawn"""
        self.selected_freq_label = label_selected
        self.redraw()
```

```
    def redraw(self):
        """Redraw the figure based on current_freq"""
        for plot_parameter in self.plot_parameters:
            if self.selected_freq_label == plot_parameter.
              frequency_label:
                self.fdtd_plot.set_data(plot_parameter.
                  fdtd_location, plot_parameter.fdtd_amplitude)
                self.bessel_plot.set_data(plot_parameter.
                  bessel_location, plot_parameter.
                    bessel_amplitude)
        self.current_ax.relim()
        self.current_ax.autoscale_view(True, True, True)
        plt.draw()


PlotParameters = namedtuple('PlotParameters',
                            ['frequency', 'frequency_label',
                             'fdtd_location', 'fdtd_amplitude',
                             'bessel_location', 'bessel_
                               amplitude'])


def load_plot_parameters():
    """ Load the arrays containing the outputs from fd3d_5_1 """

    fdtd_x_axis = np.load('fdtd_x_axis.npy')
    fdtd_amp = np.load('fdtd_amp.npy')
    bessel_x_axis = np.load('bessel_x_axis.npy')
    bessel_amp = np.load('bessel_amp.npy')
    frequencies = np.load('frequencies.npy')

    return [
        PlotParameters(
            frequency=frequency,
            frequency_label="{} MHz".format(int(frequency / 1e6)),
            fdtd_location=fdtd_x_axis,
            fdtd_amplitude=fdtd_amp[i, :],
            bessel_location=bessel_x_axis,
            bessel_amplitude=bessel_amp[i, :]
        )
        for i, frequency in enumerate(frequencies)
    ]


def plot_fig_4_7(plot_parameter):
    """Plot Fig. 5.2
    This is used to view one frequency at a time"""
```

```
    plt.rcParams['font.size'] = 12
    plt.rcParams['grid.color'] = 'gray'
    plt.rcParams['grid.linestyle'] = 'dotted'
    figure, current_ax = plt.subplots(figsize=(8, 3))

    fdtd_plot, = current_ax.plot(plot_parameter.fdtd_location,
                                 plot_parameter.fdtd_amplitude,
                                 color='k',
                                 linewidth=1)
    bessel_plot, = current_ax.plot(plot_parameter.bessel_location,
                                   plot_parameter.bessel_amplitude,
                                   'ko',
                                   mfc='none',
                                   linewidth=1)
    plt.xlabel('cm')
    plt.ylabel('Amplitude')
    plt.xticks(np.arange(-5, 10, step=5))
    plt.xlim(-9, 9)

    plt.tight_layout()
    return figure, current_ax, fdtd_plot, bessel_plot


if __name__ == '__main__':
    main()
```

6

DEEP REGIONAL HYPERTHERMIA TREATMENT PLANNING

Hyperthermia is the heating of tissues to gain a therapeutic advantage in cancer treatment and is usually done in conjunction with radiation therapy or oncology. The heating is often accomplished by radio frequency (RF) radiation. The most difficult sites to heat are deep-seated tumors like prostate, bladder, or cervix cancers. These treatments are difficult because RF energy is rapidly absorbed by human tissue. Therefore, in treating tumors more than a few centimeters below the skin surface, several applicators are used in the hope of a constructive interference pattern at the tumor site. This is a configuration usually referred to as an annular phased array (APA) (1).

In this chapter we describe the use of FDTD simulation for treatment planning in an APA. This example exploits the versatility of FDTD by incorporating concepts from signal processing as well as basic mathematical concepts like linearity and superposition to solve a complex problem. The FDTD simulation also illustrates power of the Python programing language with its ability to handle large computations and provide the graphics that display the results in a meaningful way.

Electromagnetic Simulation Using the FDTD Method with Python, Third Edition.
Jennifer E. Houle and Dennis M. Sullivan.

Published 2020 by John Wiley & Sons, Inc.

6.1 INTRODUCTION

One of the most widely used devices in deep, regional hyperthermia is the Sigma 60 applicator, which is part of the BSD-2000 hyperthermia system manufactured by BSD Medical Corporation. The Sigma 60 is a cylindrical applicator that fits around a patient undergoing treatment (Fig. 6.1). This applicator consists of eight dipole antennas evenly spaced around an annulus that is 60 cm in diameter (Fig. 6.2). The patient is positioned in the middle of the Sigma 60 and a water bolus is inflated with distilled water to give a better coupling between the antennas and the patient. The dipoles are arranged in four groups of two dipoles each, referred to as quadrants. Each quadrant is powered by its own linear class A power amplifier. All quadrants are powered at the same frequency (usually 90 MHz), but each can have a separate amplitude and phase, giving the ability to steer the power directly to the tumor site. The ability to select amplitude and phase for each quadrant presents an extremely large variety of settings that makes the selection of the optimal setting a challenging problem for the operator.

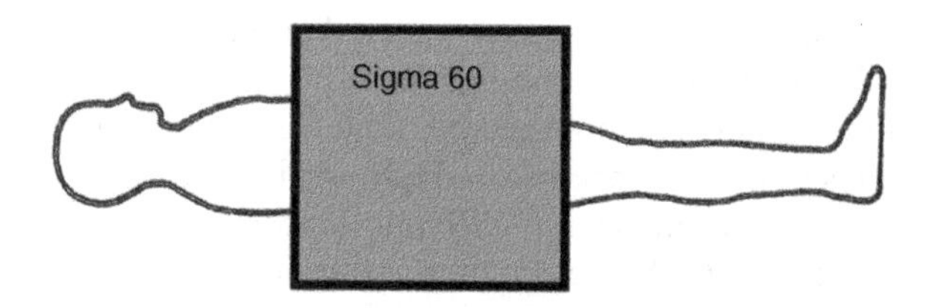

Figure 6.1 A diagram indicating the Sigma 60 applicator placed around the patient.

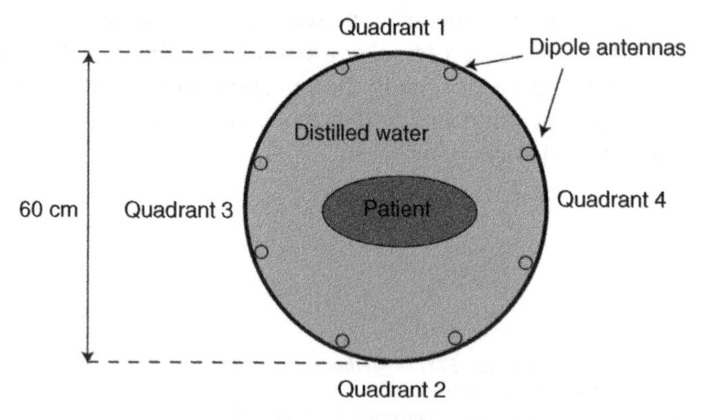

Figure 6.2 Axial view of the Sigma 60 annular phased array.

6.2 FDTD SIMULATION OF THE SIGMA 60

It has long been recognized that computer simulations can play a significant role in selecting the optimum amplitude and phase settings at each quadrant of an APA (2). Such a treatment planning program based on the FDTD method was used clinically over a period of several years (3). Recently, an updated version based on the Python programing language was presented (4).

6.2.1 Simulation of the Applicator

A diagram of the FDTD problem space with the Sigma 60 applicator is shown in Fig. 6.3. The total problem space is $80 \times 80 \times 80$ cells. Each cell is 1 cm^3. The entire problem's space is bordered by a five-cell PML.

A critical part of the FDTD simulation of the Sigma 60 is the simulation of the dipole antennas. The actual dipole antennas in the Sigma 60 are "bow tie" antennas that are fabricated on its inside wall. The FDTD program models the antennas as simple wire antennas by specifying the values GAZ = 0 at the points corresponding to the metal of the dipoles (Fig. 6.4). The input to the antennas is modeled in FDTD by setting the *Ez* value at the gap of the dipole. FDTD models only the *E* and *H* fields. Therefore, the current flowing through the wires is modeled indirectly through Ampere's law:

$$I = \oint_C \boldsymbol{H} \cdot d\boldsymbol{l}. \tag{6.1}$$

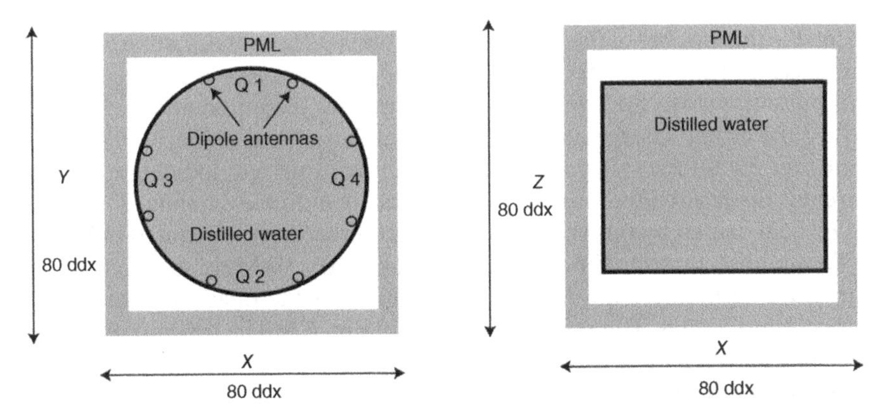

Figure 6.3 An illustration of the FDTD problem space containing the Sigma 60 applicator. The left side shows the middle XY slice and the right side shows the middle XZ slice. The total dimension is $80 \times 80 \times 80$ cells and each cell is 1 cm^3.

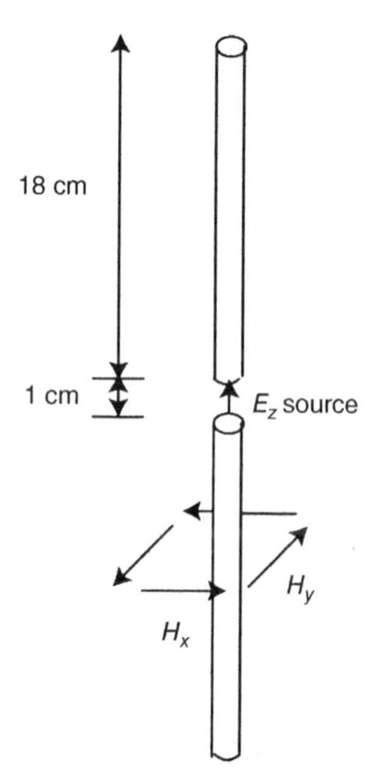

Figure 6.4 FDTD model of the dipole antennas.

In this way, we can monitor current flow in the dipole arms of quadrant one, for instance, by observing the *Hx* field directly in front of the dipole. The stimulus at the input of the dipole is

$$E_z\text{source} = \exp\left(-0.5\left(\frac{T-200}{30}\right)^2\right), \tag{6.2}$$

where T is the number of the time step. Equation (6.2) produces a narrow Gaussian pulse that causes an *Hx* pulse in the middle of the dipole at time $T = 200$. This pulse travels to the end of the dipole, reverses direction, and travels back toward the center as expected (Fig. 6.5). By taking the discrete Fourier transform over 3000 time steps, we get the pattern of Fig. 6.6, indicating that the current is distributed the way we would expect in a dipole antenna. The resulting *Ez* field caused by this radiation from quadrant one is illustrated in Fig. 6.7

To establish that the dipoles are radiating at the 90 MHz specified for the Sigma 60, we monitor the *Ez* field at a point near the quadrant one dipoles, as shown in Fig. 6.8. The results are shown in Fig. 6.9. The top half of Fig. 6.9 is the time-domain data. The source at the dipole gaps is the dashed line, which is just the Gaussian pulse of Eq. (6.2). The solid line is the *Ez* field collected at the monitor point. The bottom half of Fig. 6.9 shows the Fourier transforms of the source and the monitor field. Clearly, the monitored field is concentrated near 90 MHz.

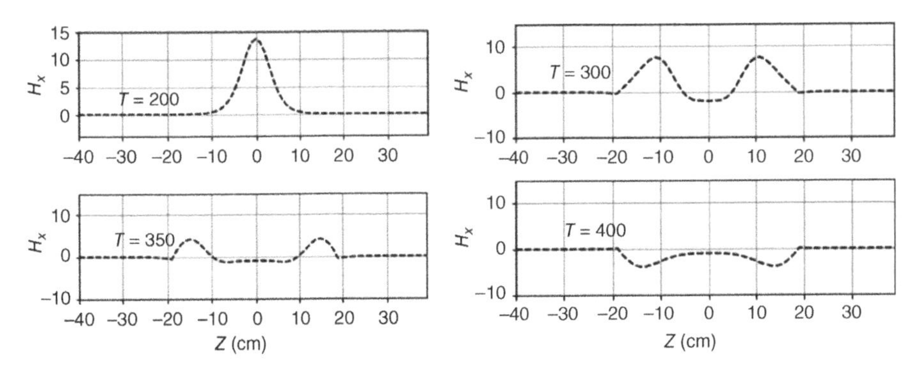

Figure 6.5 The Hx field in front of the dipole is an indication of current flow.

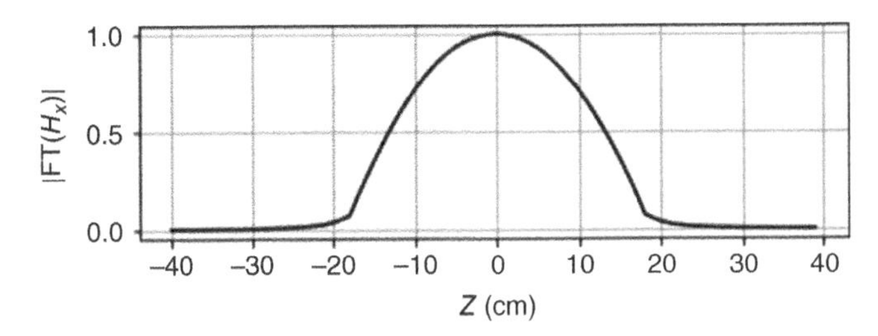

Figure 6.6 The amplitude of the Fourier transform of Hx in front of the dipole showing the type of current distribution that is expected on a dipole antenna.

6.2.2 Simulation of the Patient Model

A crucial part of the simulation is obtaining an accurate model of the patient being treated in the Sigma 60. This model must consist of the FDTD parameters that are derived from the dielectric constants and conductivities of the patient's tissues. The patient to be treated undergoes a CT scan that consists of pixel values indicating the density of their tissue. The density is used by the program **fd3d_apa.py** to identify the tissue as one of three broad classes of tissue: fat, muscle, or bone. The resulting tissue type is converted to the dielectric constant and conductivity that are needed for the FDTD simulation (5). Since the pixel values are at intervals on the order of 1 mm, many of the pixel values are averaged together before being converted to the parameters used by the FDTD program, GAX, GBX, and so on.

This process is illustrated in Fig. 6.10. A model of a female patient was taken with 150 CT scan at an interval of 1 cm (Fig. 6.10a). Figure 6.10b shows the axial view of the 75th scan. Contour illustrations of the resulting GAZ parameters are shown in Fig. 6.10c and Fig. 6.10d. Notice in the Coronal contour (Fig. 6.10c) that it is not necessary to model the entire patient, but only enough that the model extends into the PML.

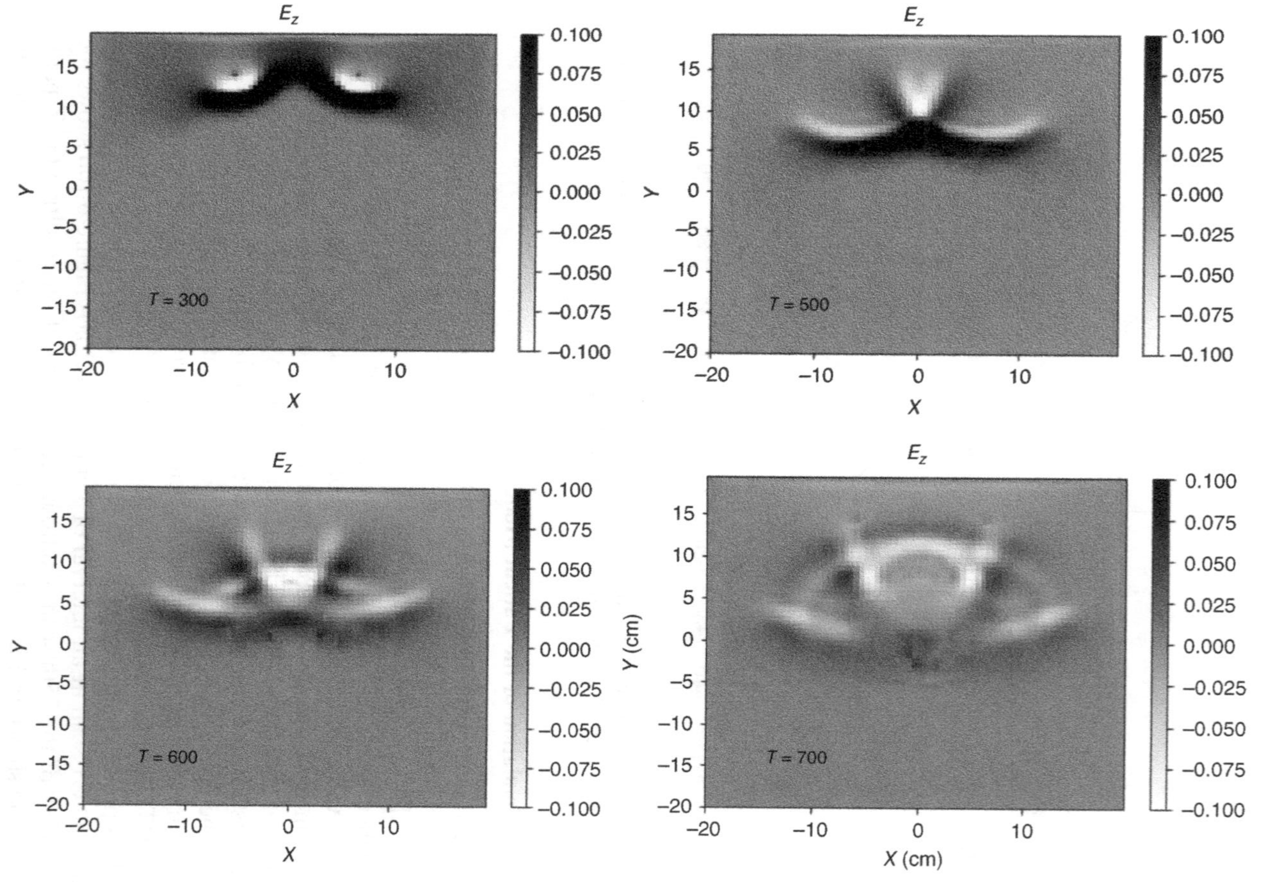

Figure 6.7 Illustration of the radiation from quadrant 1 for 300, 500, 600, and 700 time steps. By T = 700 the wave has started to interact with the patient model.

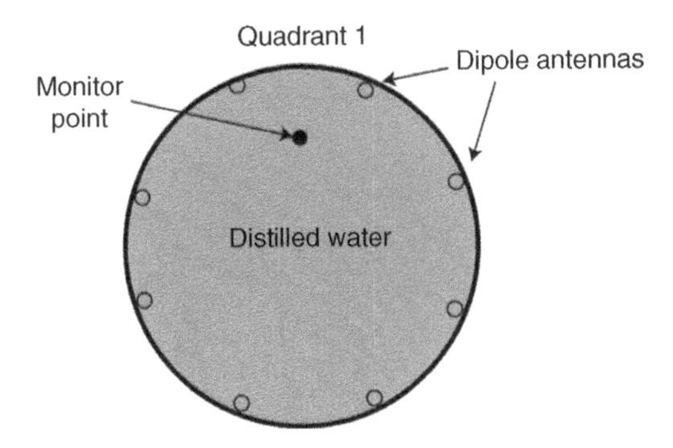

Figure 6.8 Test configuration to evaluate the frequency response of the dipole antennas.

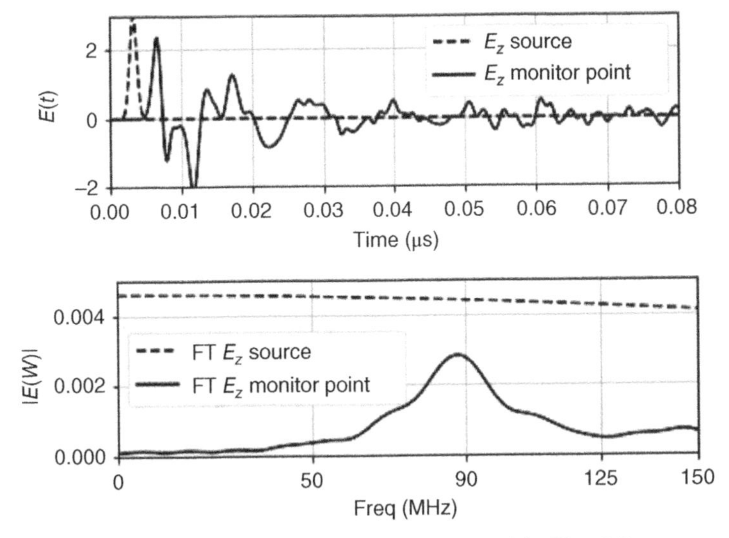

Figure 6.9 Output of the test illustrated in Fig. 6.5.

6.3 SIMULATION PROCEDURE

Once the patient model has been created, the program **fd3d_apa.py** makes four FDTD runs corresponding to the four quadrants illustrated in Fig. 6.2. Only one quadrant at a time is used as the source of EM radiations. Figure 6.7 shows the radiation from quadrant one. Even though the simulation corresponds to 90 MHz, a Gaussian pulse is used as the source on the two dipoles of quadrant one. During the simulation, a discrete Fourier transform at 90 MHz is taken at

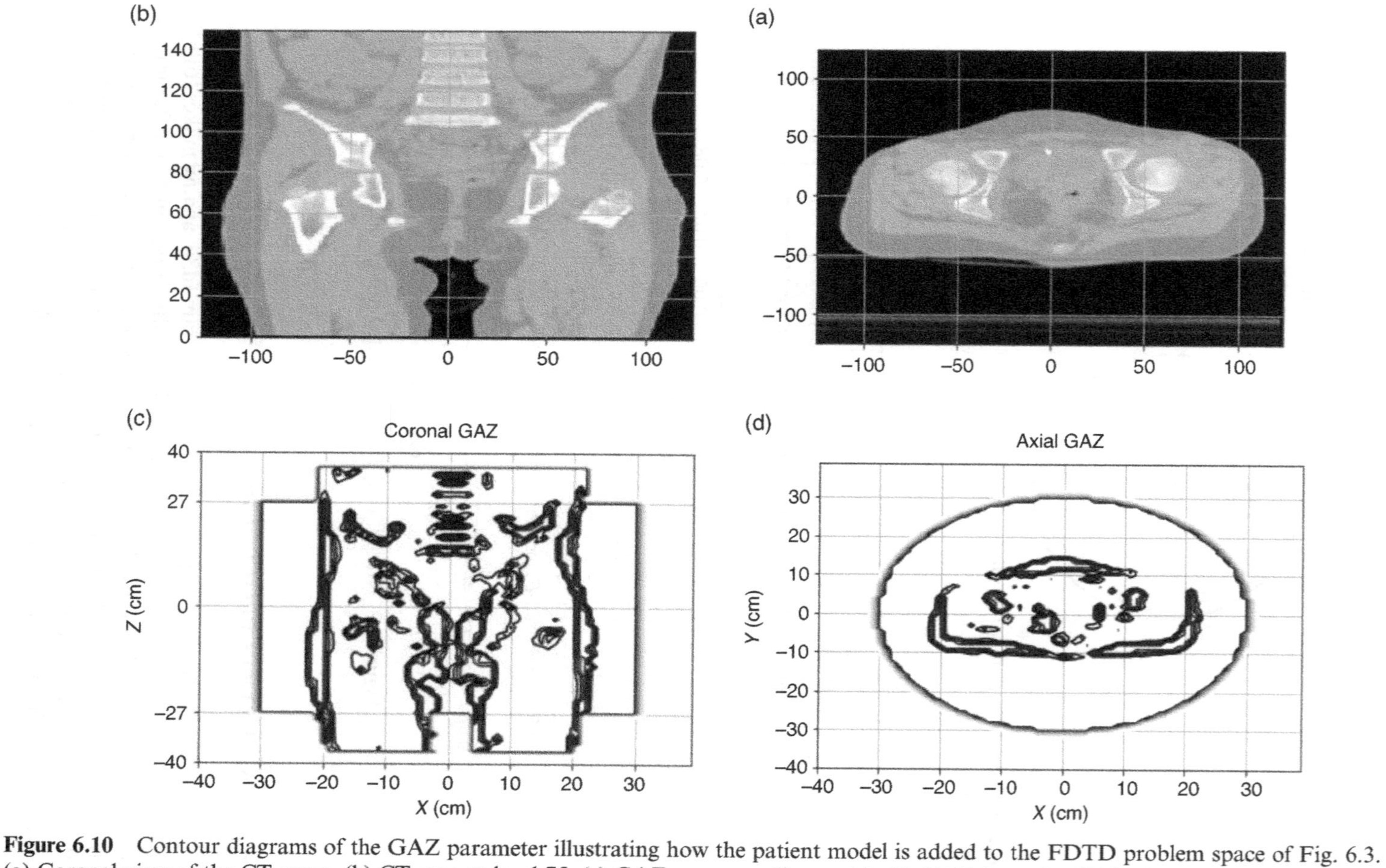

Figure 6.10 Contour diagrams of the GAZ parameter illustrating how the patient model is added to the FDTD problem space of Fig. 6.3. (a) Coronal view of the CT scans. (b) CT scan at level 75. (c) GAZ corresponding to (a). (d) GAZ from the CT in (b).

every cell within the patient model. This was explained in Chapter 2. After 5000 time steps, the resulting amplitude and phase at every cell in the model are stored in the files A1 and P1, respectively. This process is repeated for quadrants 2 through 4. The conductivity at each cell is also stored in a file. The four FDTD runs require about 10 minutes on an HP-Spectre ×360 laptop.

The parameter of interest is the specific absorption rate (SAR). This is the rate at which EM energy is being absorbed in the body tissues. It is given by

$$\mathrm{SAR}(x, y, z) = \sigma(x, y, z)\cdot E_{\mathrm{total}}^2(x, y, z), \tag{6.3}$$

where $E_{\mathrm{total}}(x, y, z)$ is the total E field at the point (x, y, z) due to the superposition of the four fields corresponding to the four quadrants. Since the dominating influence will be the E field in the Z direction, we write,

$$\boldsymbol{E}_{Z\,\mathrm{total}}(x, y, z) = \sum_{n=1}^{4} \boldsymbol{E}_{Z\,n}(x, y, z). \tag{6.4}$$

Remember that all quadrants are powered at the same frequency, in this case 90 MHz, so we can specify each E field by its amplitude and phase,

$$\boldsymbol{E}_{Z\,n}(x, y, z) = |E_{Z\,n}(x, y, z)|\angle\phi_{\mathrm{n}}. \tag{6.5}$$

Recall that we have the ability to change the magnitude and phase of the source at each quadrant. We will call these α_n and θ_n. The basic principle of linearity tells us that the total E field in the Z direction is

$$\boldsymbol{E}_{Z\,\mathrm{total}}(x, y, z) = \sum_{n=1}^{4} \alpha_1 |E_{Z\,n}(x, y, z)|\angle(\phi_{\mathrm{n}} + \theta_{\mathrm{n}}). \tag{6.6}$$

This is the value of $\boldsymbol{E}_{Z\,\mathrm{total}}(x, y, z)$ that we could use to get the total SAR using Eq. (6.4).

Rather than calculating $\boldsymbol{E}_{Z\,\mathrm{total}}(x, y, z)$ for each group of potential settings, we calculate the amplitude and phase in Eq. (6.4) for each quadrant with an input of one and no phase shift and then use superposition to calculate $\boldsymbol{E}_{Z\,\mathrm{total}}(x, y, z)$ (6). A flow chart of this process is illustrated in Fig. 6.11.

Once the four FDTD runs using **fd3d_apa.py** are complete, the operator uses the program **super_apa.py** to simulate the SAR distribution using Eq. (6.6). The operator specifies the relative amplitudes, with one normally being the maximum. However, the phases are determined indirectly by choosing a two points (ipos,jpos) relative to center for the X and Y coordinates. The resulting phases are found from the formula,

$$\theta_i = 0.1 \times (\text{frequency in MHz}) \times (dist_i - dist_{\max}). \tag{6.7}$$

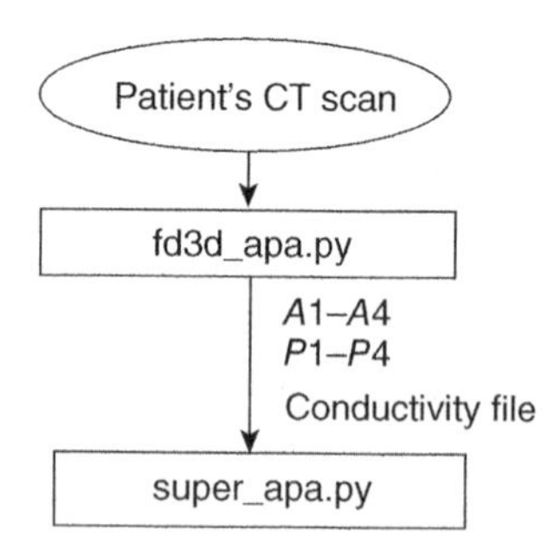

Figure 6.11 Flow chart of the treatment planning system.

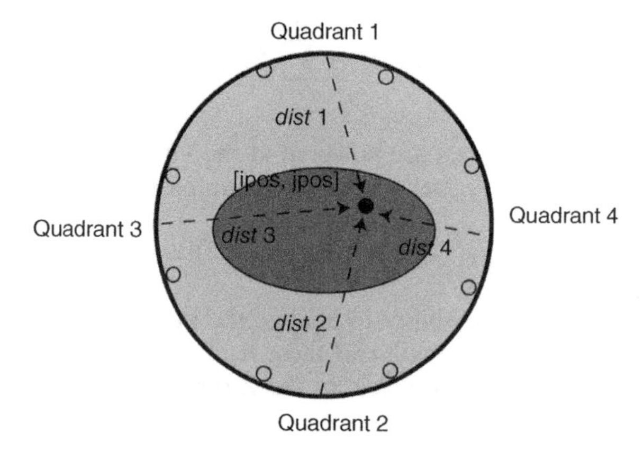

Figure 6.12 The operator specifies a target point (ipos,jpos) and the phases of the four quadrants are estimated to give maximum E field at this point.

The parameter $dist_i$ is the distance from the chosen point, (ipos,jpos), to the ith quadrant, and $dist_{\max}$ is the maximum of the four distances (Fig. 6.12).

Figure 6.13 shows four examples of the SAR distribution for various settings.

6.4 DISCUSSION

The accuracy of the treatment planning system described in this chapter was verified by measurements in a homogeneous phantom (6) and subsequently in a more realistic inhomogeneous phantom (7).

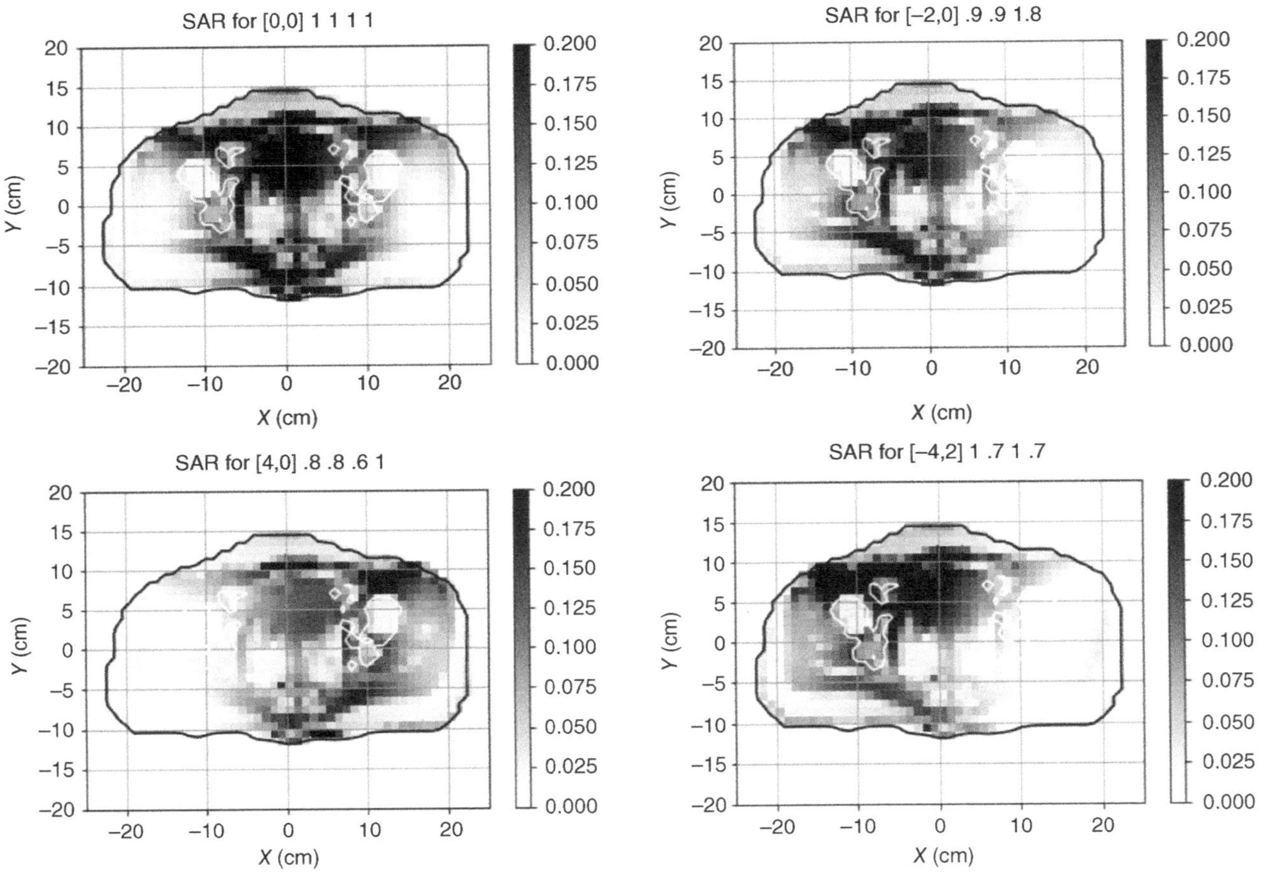

Figure 6.13 The resulting SAR distributions for four different settings. The dark regions indicate the highest anticipated SAR.

REFERENCES

1. P. F. Turner, Regional hyperthermia with an annular phase array, *IEEE Trans. Biomed. Eng.*, vol. BME-31, 1984, pp. 106–114.
2. V. Sathiaseelan, M. F. Iskander, G. C. Woward, and N. M. Bleehe, Theoretical analysis and clinical demonstration of the effect of power pattern control using the annular phased-array hyperthermia system, *IEEE Trans. Microwave Theory Tech.*, vol. MTT-17, 1986, pp. 175–185.
3. D. M. Sullivan, R. Ben-Yosef, and D. S. Kapp, The Stanford 3D hyperthermia treatment planning system—technical review and clinical summary, *Int. J. Hyperthermia*, vol. 9 (5), 1993, pp. 627–643.
4. D. M. Sullivan, J. Houle, J. Nadobny, and P. Wust, The Moscow-Berlin hyperthermia treatment planning system, 32nd Annual Meeting of the European Society for Hyperthermic Oncology, May 16–19, 2018, Berlin, Germany.
5. B. James and D. M. Sullivan, Direct use of CT scans for hyperthermia treatment planning, *IEEE Trans. Biomed. Eng.*, vol. BME-39, 1992, pp. 845–851.
6. D. M. Sullivan, Mathematical methods for treatment planning in deep regional hyperthermia, *IEEE Trans. Microwave Theory Tech.*, vol. MTT-39, 1991, pp. 864–872.
7. D. M. Sullivan, D. Beuchler, and F. A. Gibbs, Comparison of measured and simulated data in an annular phased array using an inhomogeneous phantom, *IEEE Trans. Microwave Theory Tech.*, vol. Mtt-40, 1992, pp. 600–604.

APPENDIX A

THE Z TRANSFORM

One of the most useful techniques in engineering or scientific analysis is transforming a problem from the time domain to the frequency domain (1–3). Using a Fourier or Laplace transform, differential equations are changed to algebraic equations and convolution integrals are changed to multiplication. When using discrete time-domain functions, that is, those functions defined only at points on a specific interval, we use Z transforms (4, 5).

Many engineers, particularly electrical engineers, are familiar with the Z transform, which is generally used when dealing with digital signals. While using it for FDTD simulation, we have to remember that we are implicitly sampling continuous time signals at a specific interval Δt. This presents some differences in the use of Z transforms, particularly in the convolution theorem where an extra Δt is present (7, 8).

We begin with a little mathematical background that will explain these differences.

A.1 THE SAMPLED TIME DOMAIN AND THE Z TRANSFORM

The Dirac delta function $\delta(t)$ has value at one point only, at $t = 0$, and its amplitude is such that

Electromagnetic Simulation Using the FDTD Method with Python, Third Edition.
Jennifer E. Houle and Dennis M. Sullivan.

Published 2020 by John Wiley & Sons, Inc.

$$\int_{-\infty}^{\infty} \delta(t)\mathrm{d}t = \int_{0^-}^{0^+} \delta(t)\ \mathrm{d}t = 1, \tag{A.1}$$

that is, integrating over $\delta(t)$ returns the value 1. Similarly, $\delta(t - t_0)$ is defined only at t_0 and integrating it through t_0 produces the value 1. This brings us to the *shifting theorem*,

$$\int_{-\infty}^{\infty} f(t)\delta(t - t_0)\ \mathrm{d}t = f(t_0). \tag{A.2}$$

Note that the Fourier transform of a function multiplied by a delta function is

$$\int_{-\infty}^{\infty} f(t)\delta(t - t_0)\ \mathrm{e}^{-j\omega t}\mathrm{d}t = f(t_0)\mathrm{e}^{-j\omega t_0}.$$

If we have a continuous, causal signal $x(t)$ and we sample it at an interval Δt, we get the following *discrete time* signal:

$$x[n] = \sum_{n=0}^{\infty} x(t)\delta(t - n \cdot \Delta t). \tag{A.3}$$

The Fourier transform of this discrete time function is

$$X(\omega) = x[0] + x[1]\mathrm{e}^{-j\omega \cdot \Delta t} + x[2]\mathrm{e}^{-j2\omega \cdot \Delta t} + \cdots \tag{A.4}$$

It is useful to define a parameter

$$z = \mathrm{e}^{j\omega \cdot \Delta t}. \tag{A.5}$$

The quantity $\omega \cdot \Delta t$ is radian frequency times time, so $\omega \cdot \Delta t$ is an angle. As long as the time interval Δt is chosen to be much smaller than any frequency ω associated with the problem, then $0 < \omega \cdot \Delta t < 2\pi$. The parameter z is a complex parameter with magnitude 1 and angle $\omega \cdot \Delta t$.

The Z transform of a discrete signal is defined as

$$X(z) = \sum_{n=0}^{\infty} x[n]\ z^{-n} = x[0] + x[1]z^{-1} + x[2]z^{-2} + \cdots \tag{A.6}$$

This is obviously the same as Eq. (A.4) using Eq. (A.5).

For example, let us start with the time-domain function:

$$x(t) = \mathrm{e}^{-\alpha t}u(t) \quad \alpha \geq 0.$$

If this function is sampled at an interval Δt, then it becomes

$$x[n] = \left(e^{-\alpha \cdot \Delta t}\right)^n u[n] \quad n = 0, 1, 2, \ldots$$

The Z transform of this function is

$$X(z) = \sum_{n=0}^{\infty} x[n]\, z^{-n} = \sum_{n=0}^{\infty} \left(e^{-\alpha \cdot \Delta t}\right)^n z^{-n} = \sum_{n=0}^{\infty} \left(e^{-\alpha \cdot \Delta t} z^{-1}\right)^n. \tag{A.7}$$

The following series summation will prove useful (6):

$$\sum_{n=0}^{\infty} q^n = \frac{1}{1-q} \quad \text{if} \quad |q| < 1. \tag{A.8}$$

In Eq. (A.7) as long as $e^{-\alpha \cdot \Delta t} < 1$, the absolute values of the terms in parentheses will also be less than 1. Using Eq. (A.8), Eq. (A.7) becomes

$$X(z) = \sum_{n=0}^{\infty} \left(e^{-\alpha \cdot \Delta t} z^{-1}\right)^n = \frac{1}{1 - e^{-\alpha \cdot \Delta t} z^{-1}}. \tag{A.9}$$

Note that in the limit as α goes to 0, $e^{-\alpha \cdot \Delta t}$ goes to 1, so

$$Z\{u[n]\} = \frac{1}{1 - z^{-1}}. \tag{A.10}$$

Another important Z transform is that of the decaying sinusoid:

$$\begin{aligned} f[n] &= e^{-\alpha \cdot \Delta t \cdot n} \sin(\beta \cdot \Delta t \cdot n) u[n] \\ &= \left(e^{-\alpha \cdot \Delta t \cdot n} \cdot \frac{e^{j\beta \cdot \Delta t \cdot n} - e^{-j\beta \cdot \Delta t \cdot n}}{2j}\right) u[n] \\ &= \frac{1}{2j}\left(e^{-(\alpha - j\beta) \cdot \Delta t \cdot n} - e^{-(\alpha + j\beta) \cdot \Delta t \cdot n}\right) u[n]. \end{aligned}$$

Extrapolating from Eq. (A.9) above

$$\begin{aligned} F(z) &= \frac{1}{2j}\left(\frac{1}{1 - e^{-\alpha \cdot \Delta t} e^{j\beta \cdot \Delta t} \cdot z^{-1}} - \frac{1}{1 - e^{-\alpha \cdot \Delta t} e^{-j\beta \cdot \Delta t} \cdot z^{-1}}\right) \\ &= \frac{e^{-\alpha \cdot \Delta t} \sin(\beta \cdot \Delta t) z^{-1}}{1 - 2e^{-\alpha \cdot \Delta t} \cos(\beta \cdot \Delta t) z^{-1} + e^{-2\alpha \cdot \Delta t} z^{-2}}. \end{aligned} \tag{A.11}$$

TABLE A.1 Some Z Transforms

Time Domain	Frequency Domain	Sampled Time Domain	Z Domain
$\delta(t)$	1	$\delta[n]$	$\frac{1}{\Delta t}$
$u(t)$	$\frac{1}{j\omega}$	$u[n]$	$\frac{1}{1-z^{-1}}$
$e^{-\alpha t}u(t)$	$\frac{1}{j\omega+\alpha}$	$e^{-\alpha \cdot \Delta t \cdot n}u[n]$	$\frac{1}{1-e^{-\alpha \cdot \Delta t}z^{-1}}$
$e^{-\alpha t}\sin(\beta t)u(t)$	$\frac{\beta}{(\alpha^2+\beta^2)+j\omega 2\alpha-\omega^2}$	$e^{-\alpha \cdot \Delta t \cdot n}\sin(\beta \cdot \Delta t \cdot n)u[n]$	$\frac{e^{-\alpha \cdot \Delta t}\sin(\beta \cdot \Delta t)z^{-1}}{1-2e^{-\alpha \cdot \Delta t}\cos(\beta \cdot \Delta t)z^{-1}+e^{-2\alpha \cdot \Delta t}z^{-2}}$

Finally, we will want to know the Z transform of the Dirac delta function that was defined in Eq. (A.1). The discrete time version of the delta function is $\delta[n]$. The function maintains the value 1 over an interval Δt, unlike Eq. (A.1), which implies that $\delta(t)$ is infinitely narrow. Therefore, we say that the Z transform of this function is

$$Z\{\delta[n]\} = \frac{1}{\Delta t}. \tag{A.12}$$

The $1/\Delta t$ term is necessary for mathematical consistency.

Table A.1 summarizes Eq. (A.9), Eq. (A.10), Eq. (A.11), and Eq. (A.12). These are four of the most commonly used Z transforms.

A.1.1 Delay Property

We know that the Z transform of a function $x[n]$ is given by $X(z)$ in Eq. (A.7). It is also important to know the Z transform of that same function delayed by m time steps. This is determined directly from the definition:

$$Z\{x[n-m]\} = \sum_{n=m}^{\infty} x[n-m]z^{-n}.$$

The index on the summation has been changed because we assume $x[n]$ is causal, values of $x[n-m]$ are zero until $n = m$. We change the variables to $i = n - m$, which gives

$$Z\{x[n-m]\} = \sum_{i=0}^{\infty} x[i]z^{-(i+m)} = z^{-m}\sum_{i=0}^{\infty} x[i]z^{-i} = z^{-m}X(z). \tag{A.13}$$

A.1.2 Convolution Property

Often, we are given information about a system in the frequency domain. For instance, if the transfer function of a system is $H(\omega)$ and the Fourier transform of the input is $X(\omega)$, the output in the frequency domain is

$$Y(\omega) = H(\omega)X(\omega). \tag{A.14}$$

If H and X are both causal functions, the above multiplication in the frequency domain becomes a convolution in the time domain (1):

$$y(t) = \int_0^t h(t-\tau)x(\tau)\mathrm{d}\tau.$$

If we replace $x(t)$ with the sampled time-domain function of Eq. (A.3), then

$$y(t) = \int_0^t h(t-\tau)\sum_{n=0}^{\infty} x(\tau)\delta(\tau - n\cdot\Delta t)\mathrm{d}\tau.$$

We know that delta functions only have values at the interval Δt, so we can use the shifting theorem to write the integral as a summation times Δt,

$$y(t) = \Delta t\sum_{n=0}^{\infty} h(t - n\cdot\Delta t)x(n\cdot\Delta t),$$

giving the discrete time function

$$y[m] = \Delta t\sum_{n=0}^{\infty} h[m-n]x[n].$$

We assume that t has been replaced by $\Delta t\cdot m$ and τ has been replaced by $\Delta t\cdot n$. Taking the Z transform,

$$Y(z) = \sum_{m=0}^{\infty} y[m]z^{-m} = \Delta t\cdot\sum_{m=0}^{\infty} z^{-m}\sum_{n=0}^{\infty} h[m-n]x[n],$$

and making the following change of variables

$$i = m - n, \quad m = i + n,$$

we get

TABLE A.2 Properties of Z Transforms

Property	Time Domain	Z Domain
Definition	$x[n]$	$X(z)$
Time shift	$x[n-m]$	$z^{-m}X(z)$
Convolution	$y[m] = \Delta t \sum_{n=0}^{\infty} h[m-n]x[n]$	$\Delta t \cdot H(z)X(z)$

$$\begin{aligned} Y(z) &= \Delta t \cdot \sum_{i=-n}^{\infty} z^{-i} z^{-n} \sum_{n=0}^{\infty} h[i]x[n] \\ &= \Delta t \cdot \sum_{i=0}^{\infty} h[i] z^{-i} \sum_{n=0}^{\infty} x[n] z^{-n} \\ &= \Delta t \cdot H(z)X(z). \end{aligned} \tag{A.15}$$

The summation over i was truncated to begin at zero because $h[i]$ is a causal function. The extra factor Δt in Eq. (A.15) is the difference in this version of the Z transform convolution and what is usually seen in introductory texts containing Z transforms.

Note that if we calculate the convolution of a discrete time function $x[n]$ with the delta function $\delta[n]$ in the Z domain

$$Y(z) = \Delta t \cdot X(z) \cdot \frac{1}{\Delta t} = X(z),$$

we get back the original function, as we should. These properties are summarized in Table A.2.

A.2 EXAMPLES

As an example, suppose we are to develop a computer program to calculate the convolution of the function $x(t)$ with a low-pass filter given by

$$H(\omega) = \frac{\omega_1}{j\omega + \omega_1}. \tag{A.16}$$

If $y(t)$ is the output, then

$$Y(\omega) = \frac{\omega_1}{j\omega + \omega_1} X(\omega)$$

is in the frequency domain. This filter has a cutoff frequency of $\omega_1 = 2\pi \times 10^3$ rad/s, so a sampling time of 0.01 ms should be adequate. Looking at Table A.1, the function H in the Z domain is

$$H(z) = \frac{\omega_1}{1 - e^{-\omega_1 \cdot \Delta t} z^{-1}}.$$

(Note that $\omega_1 \cdot \Delta t = 0.0628$, so $e^{-\omega_1 \cdot \Delta t} = 0.9391$.) The Z transform of the input $x(t)$ is $X(z)$, and that of the output is $Y(z)$ Therefore, by the convolution theorem, the Z transform of the output is

$$Y(z) = \Delta t \cdot \frac{\omega_1}{1 - e^{-\omega_1 \cdot \Delta t} z^{-1}} X(z). \tag{A.17}$$

We can write this as

$$\left(1 - e^{-\omega_1 \cdot \Delta t} z^{-1}\right) Y(z) = \omega_1 \cdot \Delta t \cdot X(z)$$

or

$$Y(z) = e^{-\omega_1 \cdot \Delta t} z^{-1} Y(z) + \omega_1 \cdot \Delta t \cdot X(z).$$

If we go to the sampled time domain, the above Z domain equation becomes

$$y[n] = e^{-\omega_1 \cdot \Delta t} \cdot y[n-1] + \omega_1 \cdot \Delta t \cdot x[n]. \tag{A.18}$$

This is calculated by the following Python code:

```
y[0]  = omega * dt * x[0]

for n in range(1, steps):
    y[n]  = exp(-omega * dt) * y[n - 1] + omega * dt * x[n]
```

The response to a step function is shown in Fig. A.1. Analytically, the step response to a filter of the form in Eq. (A.16) should be

$$y(t) = (1 - e^{-\omega_1 t})u(t).$$

We can see that $y[n]$ in Fig. A.1 levels off at about 1.03. The accuracy can be improved by a smaller sampling rate (see problem A.1).

Suppose our low-pass filter was a two-pole low-pass filter with the following transfer function:

$$H(\omega) = \frac{\beta}{\left(\alpha^2 + \beta^2\right) + j\alpha\omega - \omega^2}.$$

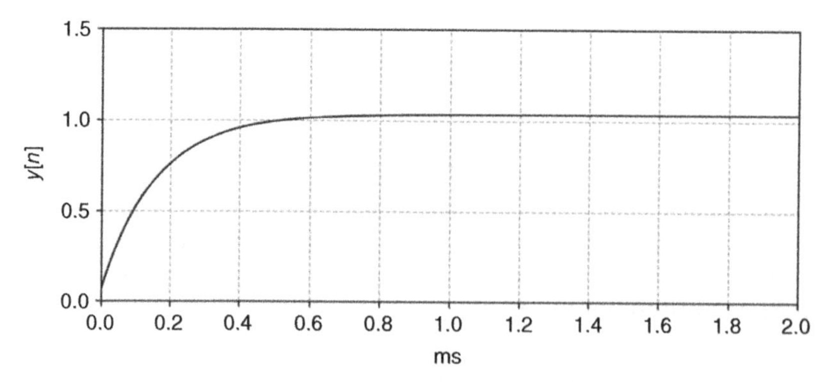

Figure A.1 The response of Eq. (A.18) to a step function input, that is, $x[n] = u[n]$.

The output for an input $X(\omega)$ in the Z domain is

$$Y(z) = \Delta t \cdot \frac{e^{-\alpha \cdot \Delta t} \sin(\beta \cdot \Delta t) z^{-1}}{1 - 2e^{-\alpha \cdot \Delta t} \cos(\beta \cdot \Delta t) z^{-1} + e^{-2\alpha \cdot \Delta t} z^{-2}} X(z).$$

The corresponding sampled time-domain function is

$$\begin{aligned} y[n] &= 2e^{-\alpha \cdot \Delta t} \cos(\beta \cdot \Delta t) \cdot y[n-1] \\ &\quad - e^{-2\alpha \cdot \Delta t} y[n-2] + e^{-\alpha \cdot \Delta t} \sin(\beta \cdot \Delta t) \cdot \Delta t \cdot x[n-1]. \end{aligned}$$

In these examples, we went directly from the frequency domain to the Z domain because the one- and two-pole low-pass filters are terms listed in Table A.1. If the frequency-domain function is a higher-order function that is not in the table, it may be possible to break the function into terms that are in the table by the method of partial fraction expansion (1).

A.3 APPROXIMATIONS IN GOING FROM THE FOURIER TO THE Z DOMAIN

In going from the frequency to the Z domain, it is difficult at times to break the frequency-domain function into separate terms that can be found in a table like Table A.1. There are methods to directly go from ω to z, but they are approximations.

It can be shown (1) that if

$$F\{f(t)\} = F(\omega)$$

then

$$F\left\{\frac{\mathrm{d}f(t)}{\mathrm{d}t}\right\} = j\omega F(\omega).$$

A derivative can be approximated by

$$\frac{\mathrm{d}f(t)}{\mathrm{d}t} \cong \frac{f(t) - f(t - \Delta t)}{\Delta t},$$

as long as Δt is small compared to how fast $f(t)$ is changing. As these may be thought of as two discrete points, we can take the Z transform:

$$Z\left\{\frac{f(t) - f(t - \Delta t)}{\Delta t}\right\} = \frac{F(z) - z^{-1}F(z)}{\Delta t} = \frac{1 - z^{-1}}{\Delta t}F(z).$$

So, at least as an approximation, we can say that $j\omega$ in the frequency domain becomes $(1 - z^{-1})/\Delta t$ in the Z domain. This is known as the backward rectangular approximation (4).

As an example, look back at the low-pass filter in Eq. (A.16). Applying the backward linear approximation gives

$$H(z) = \frac{\omega_1}{\frac{1 - z^{-1}}{\Delta t} + \omega_1} = \frac{\omega_1 \cdot \Delta t}{1 + \omega_1 \cdot \Delta t - z^{-1}} = \frac{\omega_1 \cdot \frac{\Delta t}{1 + \omega_1 \cdot \Delta t}}{1 - (1 + \omega_1 \cdot \Delta t)^{-1} z^{-1}}.$$

If we use this to calculate the convolution with of $x[n]$ with $h[n]$, we get the Z domain function similar to Eq. (A.17), which leads to the following sampled time-domain function similar to Eq. (A.18):

$$y[n] = \frac{1}{1 + \omega_1 \cdot \Delta t} y[n - 1] + \frac{\omega_1 \cdot \Delta t}{1 + \omega_1 \cdot \Delta t} x[n]. \tag{A.19}$$

Notice that as long as $\Delta t << \omega_1$, such as in the example in Section A.2,

$$\frac{\omega_1 \cdot \Delta t}{1 + \omega_1 \cdot \Delta t} \cong \omega_1 \cdot \Delta t,$$

and

$$\frac{1}{1 + \omega_1 \cdot \Delta t} \cong \mathrm{e}^{-\omega_1 \cdot \Delta t}.$$

Therefore, Eq. (A.19) is almost the same as Eq. (A.18).

The following transform is the equivalent of using a trapezoidal approximation to a derivative (4):

$$j\omega = \frac{2}{\Delta t} \cdot \frac{1 - z^{-1}}{1 + z^{-1}}. \tag{A.20}$$

Since the transform takes a linear approximation over two time steps, it is more accurate than the backward rectangular approximation. Again, starting with Eq. (A.16), by using Eq. (A.20) we get

$$\begin{aligned} H(z) &= \frac{\omega_1}{\frac{2}{\Delta t} \cdot \frac{1 - z^{-1}}{1 + z^{-1}} + \omega_1} = \frac{\omega_1 \cdot \Delta t(1 + z^{-1})}{2(1 - z^{-1}) + \omega_1 \cdot \Delta t(1 + z^{-1})} \\ &= \frac{\omega_1 \cdot \Delta t(2 + \omega_1 \cdot \Delta t)^{-1}(1 + z^{-1})}{1 - \left(\frac{1 - \omega_1 \cdot \Delta t/2}{1 + \omega_1 \cdot \Delta t/2}\right) z^{-1}} \cong \frac{\omega_1 \cdot \Delta t}{2} \cdot \frac{(1 + z^{-1})}{1 - (1 - \omega_1 \cdot \Delta t) z^{-1}}. \end{aligned} \tag{A.21}$$

This again is an approximation assuming $\Delta t < < \omega_1$ such as used in the example in Section A.2. The FDTD equation comparable to Eq. (A.19) is

$$y[n] = (1 - \omega_1 \cdot \Delta t) y[n-1] + \frac{\omega_1 \cdot \Delta t}{2}(x[n] + x[n-1]),$$

which basically indicates that the input $x[n]$ is averaged over two time steps.

PROBLEM SET A

1. Write a program that can duplicate the results of Fig. A.1. Change the time step from 0.01 to 0.001 ms. Is it more accurate?
2. Rewrite the program from problem A.1 using the backward linear approximation that results in Eq. (A.19). Do this for both time steps of 0.1 and 0.01 ms.
3. Rewrite the program from problem A.2 using the bilateral transform of Eq. (A.21). How large a time step can you use and still get fairly accurate results?

REFERENCES

1. Z. Gajic, *Linear Dynamic Systems and Signals*, Upper Saddle River, NJ: Prentice-Hall, 2003.
2. C. L. Phillips, J. M. Parr, and E. A. Riskin, *Signals, System, and Transforms*, Upper Saddle River, NJ: Prentice-Hal, 2008.

3. P. D. Chan and J. I Moliner, *Fundamentals of Signals and Systems*, Cambridge, UK: Cambridge Press, 2006.
4. A. V. Oppenheim and R. W. Schafer, *Digital Signal Processing*, Englewood Cliffs, NJ: Prentice-Hall, 1975.
5. S. K. Mitra, *Digital Signal Processing—A Computer-Based Approach*, 4th Edition, New York: McGraw-Hill, 2001.
6. E. Kreyszic, *Advanced Engineering Mathematics*, 6th Edition, New York: Wiley, 1988.
7. D. M. Sullivan, A frequency-dependent FDTD method using Z transforms, *IEEE Trans. Antennas Propag.*, vol. 40, Oct. 1992, pp. 1223–1230.
8. D. M. Sullivan, Z transform theory and the FDTD method, *IEEE Trans. Antennas Propag.*, vol. 44, Oct. 1999, pp. 28–34.

APPENDIX B

ANALYTIC SOLUTION TO CALCULATING THE ELECTRIC FIELD

In Chapters 3 and 4, the results of the FDTD simulation were compared with an analytic solution. The program used to when examining a dielectric sphere in Chapter 4 is presented here. This program uses the Hemholtz equation to calculate the electric field by making use of Bessel and Legendre functions as described in chapter 6 of Harrington (1).

```
""" bessel_3d.py: 3D Analytic Solution

Uses Bessel functions and Legendre functions to calculate the
E field in a layered dielectric sphere to solve the Hemholtz
equation as shown in chapter 6 in Harrington


"""

import cmath
from collections import namedtuple
from math import pi, sqrt, acos, atan2, sin, cos
```

Electromagnetic Simulation Using the FDTD Method with Python, Third Edition.
Jennifer E. Houle and Dennis M. Sullivan.

Published 2020 by John Wiley & Sons, Inc.

```
import matplotlib.pyplot as plt
import numpy as np

Layer = namedtuple('Layer', ('radius',
                              'sigma',
                              'epsilon'))

Sphere = namedtuple('Sphere', ('radius',
                                'k',
                                'y',
                                'sigma',
                                'epsilon_r'))

Coefficients = namedtuple('Bessel', ('bn',
                                      'cn',
                                      'dn',
                                      'en'))


def main(n_partitions,
         n_points,
         center_location,
         layers,
         freq,
         cell_size):
    """

    Args:
        n_partitions: number of partitions chosen for the simulation
        n_points: number of points at which the E field is calculated
        center_location: center location of the sphere
        layers: a list of the layers (each layer has a radius,
            sigma, epsilon)
        freq: frequency of the simulation
        cell_size: cell size in meters

    Returns:
        ez_saved: array of the e-field in the z direction for n_points

    """
    n_layers = len(layers)
    omega = 2 * pi * freq

    sphere = setup_sphere(layers, omega)

    bessel_coefficients = find_bessel_coefficients(n_partitions,
                                                   n_layers,
                                                   sphere)
```

```
    ez_saved = calculate_e(n_points, center_location, sphere, n_layers,
                           n_partitions, bessel_coefficients, cell_size,
                                  omega)
    return ez_saved


def setup_sphere(layers, omega):
    """

    This function creates Sphere which has arrays of length number of
    layers for each of the parameters of the of the sphere.

    Args:
        layers: a list of the layers (each layer has a radius,
            sigma, epsilon)
        omega: 2 * pi * frequency

    Returns:
       namedtuple Sphere (arrays for each radius, epsilon_r, sigma, k,
       and y; these contain information about each layer of the sphere)

    """
    n_layers = len(layers)

    k = np.zeros(n_layers + 1, dtype='complex')
    y = np.zeros(n_layers + 1, dtype='complex')

    epsilon0 = 8.854e-12

    radius = np.zeros(n_layers + 1)
    epsilon_r = np.zeros(n_layers + 1)
    sigma = np.zeros(n_layers + 1)

    for idx, layer in enumerate(layers):
        complement_layer = n_layers - 1 - idx

        radius[complement_layer] = layer.radius
        epsilon_r[complement_layer] = layer.epsilon
        sigma[complement_layer] = layer.sigma

    k[n_layers] = (1 + 0j) * omega / 2.99793e8
    k[:-1] = np.sqrt(
        (epsilon_r[:-1] - 1j * sigma[:-1] /
         (omega * epsilon0)) /
        (1 + 0j)) * k[n_layers]
```

```
    y[n_layers] = 1j * omega * epsilon0 * (1 + 0j)
    y[:-1] = y[n_layers] * (
       (epsilon_r[:-1] - 1j * sigma[:-1] / (omega * epsilon0)) / (1 + 0j))

    return Sphere(
        radius=radius,
        epsilon_r=epsilon_r,
        sigma=sigma,
        k=k,
        y=y
    )


def find_bessel_coefficients(n_partitions, n_layers, sphere):
    """
    This calculates the Bessel coefficients from chapter 6 in Harrington
    (see problem 6-25). The Bessel and Hankel functions returned from
    the function 'bessel' are labeled *_k1 and *_k2 to denote whether they
    were calculated with k*r of the layer associated with the current
    radius (*_k1) or the k*r of the next outermost layer (*_k2).

    Args:
       n_partitions: number of partitions chosen for the calculation
       n_layers: number of layers in the sphere
       sphere: namedtuple Sphere (arrays for each radius, epsilon_r,
          sigma, k, and y; these contain information about each layer
          of the sphere)

    Returns:
       Coefficients: named tuple consisting of the four
          Bessel coefficients

    """
    bn=np.zeros((n_layers+1,n_partitions),dtype='complex')
    cn=np.zeros((n_layers+1,n_partitions),dtype='complex')
    dn=np.zeros((n_layers+1,n_partitions),dtype='complex')
    en=np.zeros((n_layers+1,n_partitions),dtype='complex')

    bn[0, :] = 1 + 0j
    cn[0, :] = 0 + 0j
    dn[0, :] = 1 + 0j
    en[0, :] = 0 + 0j

    for layer in range(0, n_layers):
        radius = sphere.radius[layer]
        k1 = sphere.k[layer]
        k2 = sphere.k[layer + 1]
```

```
        jn_k1, hn_k1 = bessel(k1 * radius, n_partitions)
        jn_k2, hn_k2 = bessel(k2 * radius, n_partitions)

        for partition in range(0, n_partitions):
            jn_d_k1 = -jn_k1[partition + 1] + (
                (partition + 2) / (k1 * radius)) * \
                      jn_k1[partition]

            hn_d_k1 = -hn_k1[partition + 1] + (
                (partition + 2) / (k1 * radius)) * \
                      hn_k1[partition]

            jn_d_k2=-jn_k2[partition+1]+(partition+2)/(
                k2 * radius) * jn_k2[partition]

            hn_d_k2=-hn_k2[partition+1]+(partition+2)/(
                k2 * radius) * hn_k2[partition]

            bn[layer+1,partition]=calc_coefficient(layer,partition,
                                                   bn,cn,-k2/k1,
                                                   hn_d_k2, jn_k1,
                                                   hn_k2, jn_d_k1,
                                                   hn_k1, hn_d_k1)
            cn[layer + 1, partition] = calc_coefficient(layer, partition,
                                                   bn, cn, k2 / k1,
                                                  -jn_d_k2, jn_k1,
                                                   jn_k2, jn_d_k1,
                                                   hn_k1, hn_d_k1)

            dn[layer + 1, partition] = calc_coefficient(layer, partition,
                                                   dn,en,-k1/k2,
                                                   hn_d_k2, jn_k1,
                                                   hn_k2, jn_d_k1,
                                                   hn_k1, hn_d_k1)
            en[layer + 1, partition] = calc_coefficient(layer, partition,
                                                   dn, en, k1 / k2,
                                                  -jn_d_k2, jn_k1,
                                                   jn_k2, jn_d_k1,
                                                   hn_k1, hn_d_k1)

    for partition in range(0, n_partitions):
        an = (1j ** (-(partition + 1))) * (2.0 * (partition + 1) + 1) / (
            (partition + 1) * (partition + 2))

        cn1 = an / bn[n_layers, partition]
        cn2 = an / dn[n_layers, partition]
```

```
        bn[:, partition] = bn[:, partition] * cn1
        cn[:, partition] = cn[:, partition] * cn1
        dn[:, partition] = dn[:, partition] * cn2
        en[:, partition] = en[:, partition] * cn2

    return Coefficients(bn, cn, dn, en)


def calc_coefficient(layer, partition, c1, c2, k_ratio,
                     func1, func2, func3, func4, func5, func6):
    """ Calculate a coefficient at a given partition / layer """
    return (
        c1[layer, partition] * 1j * (func1 * func2[partition] +
                                     k_ratio*func3[partition]*
                                     func4)+
        c2[layer, partition] * 1j * (func1 * func5[partition] +
                                     k_ratio*func3[partition]*
                                     func6))


def bessel(kr, n_partitions):
    """
    This calculates some of the Bessel functions and Hankel functions based
    off chapter 6 / appendix D in Harrington.

    Args:
        kr: k * r at the current layer
        n_partitions: number of partitions chosen

    Returns:
        jn: Bessel functions of the first kind
        hn: Hankel functions of the first kind

    """
    m = n_partitions + 3
    jn = np.zeros(m + 1, dtype='complex')
    jn[m - 1] = 1 + 0j

    for partition in range(m - 1, -1, -1):
        jn[partition - 1] = ((2.0 * (partition + 1) + 1) * jn[
            partition] / kr) - jn[partition + 1]

    k = ((cmath.sin(kr) / (kr ** 2)) - cmath.cos(kr) / kr) / jn[0]

    jn[:n_partitions + 1] = jn[:n_partitions + 1] * k
    jn = kr * jn
```

```
    hn = np.zeros(n_partitions + 1, dtype='complex')
    hn[0] = -cmath.cos(kr) / (kr ** 2) - cmath.sin(kr) / kr
    hn[1] = ((1 / kr) - 3.0 / (kr ** 3)) * cmath.cos(kr) \
            - 3.0 / (kr ** 2) * cmath.sin(kr)

    for partition in range(1, n_partitions):
        hn[partition + 1] = ((2 * (partition + 1) + 1)
                              * hn[partition] / kr) - hn[partition - 1]

    hn = jn[:n_partitions + 1] - 1j * kr * hn

    return jn[:n_partitions + 1], hn


def calculate_e(n_points, center_location, sphere, n_layers,
                n_partitions, bessel_coeff, cell_size, omega):
    """
    Iterates through each location to calculate the electric
      field in the z
    direction

    Args:
        n_points: number of points at which the E field is calculated
        center_location: center location of the sphere
        sphere: namedtuple Sphere (arrays for each radius, epsilon_r,
           sigma, k, and y; these contain information about each layer
           of the sphere)
        n_layers: number of layers in the sphere
        n_partitions: number of partitions chosen for the simulation
        bessel_coeff: named tuple consisting of the four Bessel
          coefficients
        cell_size: cell size in meters

    Returns:
        An array of ez at every location

    """
    ez_saved = np.zeros(n_points)

    for point in range(n_points - 1, -1, -1):
        current_r, theta, phi, cond, layer = \
            find_location_parameters(point, center_location,
              cell_size, n_layers, sphere)

       sum_r, sum_theta_1, sum_theta_2, sum_phi_1, sum_phi_2 = \
          calculate_sums(n_partitions, layer, current_r, sphere,
                         bessel_coeff, omega, n_layers, theta)
```

```
        e_r = cos(phi) * sin(theta) / (sphere.y[layer] *
                          current_r ** 2) * sum_r
        e_theta = cos(phi) / current_r * (sum_theta_2 - sum_theta_1)
        e_phi = sin(phi) / current_r * (sum_phi_1 - sum_phi_2)

      ez_saved[point] = abs(convert_to_cartesian(e_r, e_theta, e_phi,
                                                  theta, phi))

    return ez_saved


def find_location_parameters(point, center_location, cell_size,
  n_layers,
                                 sphere):
    """

    Args:
        point: the cell at which the calculation is occurring
        center_location: center location of the sphere
        cell_size: cell size in meters
        n_layers: number of layers in the sphere
        sphere: namedtuple Sphere (arrays for each radius, epsilon_r,
           sigma, k, and y; these contain information about each layer
           of the sphere)

    Returns:
       current_r: distance of the current location to the center
           of the sphere
        theta: angle theta at current point
        phi: angle phi at current point
        cond: conductivity of the current layer
        layer: the layer at the given radius

    """
    xc = 0.0
    zc = 0.005
    yc = ((point + 1) - center_location) * cell_size
    current_r = sqrt(xc ** 2 + yc ** 2 + zc ** 2)
    theta = pi - acos(yc / current_r)
    phi = atan2(xc, zc)

    cond = find_current_cond(n_layers, current_r, sphere)
    layer = find_current_layer(n_layers, current_r, sphere)

    return current_r, theta, phi, cond, layer
```

```
def find_current_cond(n_layers, rad, sphere):
    """ Returns the conductivity at the input radius

    Args:
        n_layers: number of layers in the sphere
        rad: distance of the current location to the center of the sphere
        sphere: namedtuple Sphere (arrays for each radius, epsilon_r,
            sigma, k, and y; these contain information about each layer
            of the sphere)

    Returns:
        Conductivity of the layer at the current location

    """
    cond = sphere.sigma[0]

    for layer in range(0, n_layers):
        if rad >= sphere.radius[layer]:
            cond = sphere.sigma[layer + 1]

    return cond


def find_current_layer(n_layers, rad, sphere):
    """ Returns the layer at the input radius

    Args:
        n_layers: number of layers in the sphere
        rad: distance of the current location to the center of the sphere
        sphere: namedtuple Sphere (arrays for each radius, epsilon_r,
            sigma, k, and y; these contain information about each layer
            of the sphere)

    Returns:
        the layer at the given radius

    """
    for layer, this_radius in enumerate(sphere.radius[:n_layers]):
        if rad <= this_radius:
            return layer
    return n_layers


def calculate_sums(n_partitions, layer, current_r, sphere,
                   bessel_coeff, omega, n_layers, theta):
    """
```

```
This function calculated five sums used to calculate the electric
field in
the E_r, E_theta, and E_phi directions (see eq. 6-26, problem 6-25 in
Harrington) using the Bessel, Hankel, and Legendre functions.

Args:
    n_partitions: number of partitions chosen for the simulation
    layer: the layer at the given radius
    current_r: distance of the current location to the center
        of the sphere
    sphere: namedtuple Sphere (arrays for each radius, epsilon_r,
        sigma, k, and y; these contain information about each layer
        of the sphere)
    bessel_coeff: named tuple consisting of the four Bessel
    coefficients
    omega: 2 * pi * frequency
    n_layers: number of layers in the sphere
    theta: angle theta at current point

Returns:
   sum_r, sum_theta_1, sum_theta_2, sum_phi_1, sum_phi_2, each of
   which is used in the electric field calculations

"""
jn, hn = bessel(sphere.k[layer] * current_r, n_partitions)
pn, pn_d = legendre(theta, n_partitions)

sum_r = sum_theta_1 = sum_theta_2 = sum_phi_1 = sum_phi_2 = 0 + 0j

mu0 = 4 * pi * 1.0e-7

for partition in range(0, n_partitions):
    jn_d = -jn[partition + 1] + (partition + 2) / (
        sphere.k[layer] * current_r) * jn[partition]
    hn_d = -hn[partition + 1] + (partition + 2) / (
        sphere.k[layer] * current_r) * hn[partition]

    sum_r += ((partition + 1) * (partition + 2)) / (omega * mu0) * (
        bessel_coeff.bn[layer, partition] * jn[partition] +
        bessel_coeff.cn[layer, partition] * hn[partition]) \
            * pn[partition]

    sum_theta_1 += 1 / sphere.k[n_layers] * (
        bessel_coeff.dn[layer, partition] * jn[partition] +
        bessel_coeff.en[layer, partition] * hn[partition]) \
                  * pn[partition]
```

```
        sum_theta_2 += sphere.k[layer] / \
                       (sphere.y[layer] * omega * mu0) * (
                        bessel_coeff.bn[layer, partition] * jn_d +
                       bessel_coeff.cn[layer, partition] * hn_d) \
                       * pn_d[partition]

        sum_phi_1 += 1 / sphere.k[n_layers] * (
            bessel_coeff.dn[layer, partition] * jn[partition] +
           bessel_coeff.en[layer, partition] * hn[partition]) \
                     * pn_d[partition]

        sum_phi_2 += sphere.k[layer] / \
                     (sphere.y[layer] * omega * mu0) * (
                     bessel_coeff.bn[layer, partition] * jn_d +
                     bessel_coeff.cn[layer, partition] * hn_d) \
                     * pn[partition]

    return sum_r, sum_theta_1, sum_theta_2, sum_phi_1, sum_phi_2


def legendre(theta, n_partitions):
    """

    This calculates some of the lower-order Legendre polynomials
      based off
    chapter 6 / appendix E in Harrington.

    Args:
        theta: angle theta at current point
        n_partitions: number of partitions

    Returns:
        pn: Legendre polynomials
        pn_d: derivative of Legendre polynomials

    """
    pn = np.zeros(n_partitions)
    pn[0] = 1.0
    pn[1] = 3.0 * cos(theta)

    for partition in range(2, n_partitions):
        pn[partition] = \
            ((2 * partition + 1) * cos(theta) * pn[partition - 1]
              - (partition + 1) * pn[partition - 2]) / partition

    pn2 = np.zeros(n_partitions)
    pn2[0] = 0
    pn2[1] = 3.0 * (sin(theta) ** 2)
    pn2[2] = 15.0 * cos(theta) * (sin(theta) ** 2)
```

```
    for partition in range(3, n_partitions):
        pn2[partition] = \
            ((2 * partition + 1) * cos(theta) * pn2[partition - 1] -
             (partition + 2) * pn2[partition - 2]) / (partition - 1)
    pn_d = -pn2 + cos(theta) * pn

    return pn, pn_d


def convert_to_cartesian(e_r, e_theta, e_phi, theta, phi):
    return e_r * sin(theta) * cos(phi) + e_theta * cos(theta) \
        * cos(phi) - e_phi * sin(phi)


if __name__ == '__main__':
    ez_saved = main(
        n_partitions=8,
        n_points=20,
        center_location=11,
        layers=[Layer(radius=0.1, sigma=0.3, epsilon=30)],
        freq=500e6,
        cell_size=0.01
    )

    plt.plot(ez_saved[::-1])
    plt.show()
```

REFERENCE

1. R. F. Harrington, *Time-Harmonic Electromagnetic Fields*, 2nd Edition, New York: Wiley–IEEE Press, 2001.

INDEX

Electromagnetic Simulation Using the FDTD Method with Python, Third Edition.
Jennifer E. Houle and Dennis M. Sullivan.

Published 2020 by John Wiley & Sons, Inc.